What people are saying about

Clearing a Way

I really like this. This is the most interesting model I have come across in years.
Sarah Knox, Emeritus Professor of Biophysics, West Virginia

This all looks very interesting, and we shall notify Galileo Commission members as well.
David Lorimer, Programme Director for Scientific and Medical Network

Clearing a Way

Unveiling the Mental Tricks That Hide Reality

Previous Books

Lost and Then Found: Turning Life's Disappointments into Hidden Treasures. Paternoster Press, 1999. ISBN 0-85364-966-9

Emotional Logic: Harnessing Your Emotions into Inner Strength. Jointly written with Dr Marian Langsford. Hammersmith Health Books, 2021. ISBN 978-1-78161-182-1

Shelly and Friends. Integrated series of seven illustrated books for parents and children aged 5–9. Written jointly with Sarah Lakey and Lisa Chaffer. Illustrated by Bradley Goodwin. Emotional Logic Publishing, 2018. ISBN 9-78196-4644-4-5

Individual books:

Shelly in Shock. ISBN 978-0-9954804-0-7
Ollie in Denial. ISBN 978-0-9954804-1-4
Reggie Gets Angry. ISBN 978-0-9954804-2-1
Zora Feels Guilty. ISBN 978-0-9954804-3-8
Monty Risks a Bargain. ISBN 978-0-9954804-4-5
Esther Feels Empty. ISBN 978-0-9954804-5-2
Gemmie Accepts the Long View. ISBN 978-0-9954804-6-9

Clearing a Way

Unveiling the Mental Tricks That Hide Reality

Trevor Griffiths

BOOKS

Winchester, UK
Washington, USA

JOHN HUNT PUBLISHING

First published by O-Books, 2024
O-Books is an imprint of John Hunt Publishing Ltd., 3 East St., Alresford, Hampshire SO24 9EE, UK
office@jhpbooks.com
www.johnhuntpublishing.com
www.o-books.com

For distributor details and how to order please visit the 'Ordering' section on our website.

Text copyright: Trevor Griffiths 2022

ISBN: 978 1 80341 289 4
978 1 80341 290 0 (ebook)
Library of Congress Control Number: 2022942996

All rights reserved. Except for brief quotations in critical articles or reviews, no part of this book may be reproduced in any manner without prior written permission from the publishers.

The rights of Trevor Griffiths as author have been asserted in accordance with the Copyright, Designs and Patents Act 1988.

A CIP catalogue record for this book is available from the British Library.

Design: Lapiz Digital Services

UK: Printed and bound by CPI Group (UK) Ltd, Croydon, CR0 4YY
Printed in North America by CPI GPS partners

The author of this book does not dispense medical advice or prescribe the use of any technique as a form of treatment for physical, emotional, or medical problems without the advice of a physician, either directly or indirectly. The intent of the author is only to offer information of a general nature to help you in your quest for emotional and spiritual well-being. In the event you use any of the information in this book for yourself, which is your constitutional right, the author and the publisher assume no responsibility for your actions.

We operate a distinctive and ethical publishing philosophy in all areas of our business, from our global network of authors to production and worldwide distribution.

Contents

The decades of thought and work brought together here are all dedicated to my wife Marian, our daughters Abigail and Tamsin, and our grandchildren Eleanor, Toby, Isabel, and Adeline, all of whom remind me how important it is to keep my feet on the ground and play, even while my head searches the clouds.

List of Figures and Tables

(Colour versions of these figures can be downloaded as a PDF file from www.relatedness.net/)

Preface

I was an unusual student at Oxford University, getting around on a Triumph 350 Twin motorbike with panniers rather than the traditional pushbike. In 1969 and the early 1970s I was studying medicine, neuroscience, and cell populations during those iconic years of civilised exploration from which the New Science was later to be born, more of which later. The buzz then among the teachers was all about 'form and function'—how function determined form, and how form would then constrain the function. It was the age of hope, when Buckminster Fuller was leading the revolution in architecture and futurism along the same lines, aiming to shape society to come. But from the very outset in that social atmosphere I thought, *There is something missing from this schema of life and social science.* It was relatedness. Form, function, and relatedness, to my mind, always did give a fuller picture of life experience, and of my hope for the future; but I had to buckle in and show that I understood life as merely body and mind.

My assurance that something was missing arose from having several mysterious experiences over the years of my then short life. They had been peak events that had broken the scientific rules of space and time, two of which I shall now briefly relate to you as they relate to my life with a beloved motorbike. The first arose when I was exploring a way to live intuitively with non-religious prayer, or mindfully as it might be called now. I had felt inspired to write a poem, which came out thus:

Angles turn
and lines forgive
criss-crossing of the mind.
For here in the world there is beauty

such as always found
between the seconds of your mind.

Having completed and written this inspirational thought to my satisfaction, I felt a sudden urge to take it for a ride on my motorbike. Perhaps it was intended to be given to someone who needed to read it? So I put the poem securely in an inner pocket behind two layers of jacket and set out intuitively through the centre of the city, leading out to a semi-industrial area where I felt a prompt to stop. I stood by the bike for several minutes, but no-one passed by, and so I thought perhaps I should hold the poem in my hand, prepared to give it to someone who might approach equally inspirationally. I searched, and searched, then took off my jackets and explored every pocket. There was no poem to be found. I returned to my room and searched everywhere. No poem. In this book, written nearly 50 years later, you may see that this poem accurately describes the model of implicate order to life, beyond space and time and substance, that I started to develop a mere 30 years after that thought had been written. It had been written then to deliver to you now.

I later had a dream that predicted an event so accurately that I could recognise the situation as it arose, and choose to do something counterintuitive that saved my life. In this precognitive dream I had seen my death, knocked sideways off the motorbike. Every day of my life now I owe to having taken note of the information in that dream. As a result of this non-linear event in time, having pulled in safely to the kerb I was able to watch a driver three cars in front of me make a disastrous and totally unpredictable spin in the middle of the road that would definitely have knocked me off sideways had I continued driving, and would surely have killed me. It happened at an unusually offset junction with traffic lights on the Summertown Road on the way out from the centre of Oxford. You can visit the place even now, and marvel at life's mysteriously deep structure and order as I did when returning on a nostalgia visit one year.

It bothers me not that most intelligent people respond to this account by claiming it was coincidence. I listen to their reasoning carefully and respectfully, learning over the following years from their arguments how their imaginations have been veiled in different ways by educational fashions in different cultures that guide their reasoning. There is more than body and mind. There is also the quality of relatedness that connected my inner heart with that bizarre driver's, our connection emerging in a dream a month before that incident. Having considered this for decades, I now believe that the dream arose from a clash of our personal values which, at a human inner heart level, are embedded in movements of the implicate order of life, prior to emergent space, time, and substance.

How could this be?

It all comes down to our powers of choice, which we shall need to exercise wisely in the coming decades as our environments crumble and our societies are thrown into turmoil. What follows is an ecological account of the human being, leaving behind the material and psychological views that have shaped the past three centuries of exploratory thought about body and mind. A vast, totally uncontrolled Enlightenment Science experiment has stripped life of its social meaning over the last 300 years, the effects of which now need to be untangled. The explanation that follows is a paradigm shift of vision, thought, and emotive action, which has the power to shape the future around recognising and naming our own and each other's personal values.

Acknowledgements

There have been many, many people along the way of discovery to whom I am indebted for their thoughtful influence, possibly unknown as such at the time. I think, for example, of one mathematics student at Oxford who, like me, believed he had found a way to describe the source of consciousness, and when we met in his rooms I, a medical and neuroscience student, found I could make no sense whatsoever of his formulae and symbols. We looked at each other, aware of the language divide between us that was nevertheless joined at our depths of value and hope. A memorable experience.

But there are many more with whom there have been active and deeply formative conversations over the decades. Principally Curtis Alcock, whose shared interest in chaos theory and complex adaptive systems, combined with his technical graphics skills, has produced this book's diagrams, which make accessible the ideas that have resonated between us, and still do. Trisha Horgan is the senior psychiatric nurse and group analytic therapist with whom Emotional Logic developed as a lifelong learning alternative to therapy. Her trinitarian spirituality has also helpfully kept the focus on responsive relatedness as a component of health, despite all the fashions of psychology that come and go. Marian, my wife, must figure in this roll of honour, whose earthed common sense and practical focus have grasped the meaning of these ideas about core human nature and explored how to teach it. And Neil Collins, my nonconformist pastor and friend for many decades, has helped to unveil where theology has obscured a living truth rather than revealed it, focusing again on how truth emerges among the fellowship and networks of people's everyday lives.

My thanks also go to Andrew Newcombe and Ricardo Pereira from Tooltech, who explored with me using computer-assisted design and 3D printing how the principles of organisation explained here can be manifest in physical structures and relational movement. I remember the sense of awe spreading as we saw unfolding on the screens new depths to the potential holomovement that shapes the cosmos. Along with them must be mentioned Colette Campbell-Jones, psychologist and fine artist, without whom conversations about Karen Barad, Bergson, and Alfred North Whitehead would not have happened, keeping the process view on life in tune with lived aesthetics. Likewise Prof. Sarah Knox has helped to explore how triquetral cosmology connects with biophysics and the world of scientific health research. This text is enriched by them all.

I would like to also mention how helpful the discovery has been of David Lorimer, Bernard Carr, Paul Filmore, and Paul Kieniewicz's leadership of the respectful conversations among widely diverse people in the Scientific and Medical Network forums and the Galileo Commission, exploring the interface of Enlightenment Science with consciousness. Diversification in unity is a central theme of the following text. To find it lived out so constructively has been a deep encouragement.

Finally, may I acknowledge the professionalism and responsive guidance of too many people to mention by name at the John Hunt publishing house, but who have picked up the theme and said, "This book *must* be published."

Introduction

Everything in this book orbits around one purpose. It is how to strengthen a person's inner heart to make life-enhancing choices in difficult situations, not death-enhancing ones.

Since the 1970s, a *New Science* of adaptive systems and emergent order has appeared on the scene. This dynamic way to understand life and matter has overtaken the *Enlightenment Science* of the previous 300 years, giving a more life-enhancing context to the value of that earlier method of study. In turbulent times, moving with the New Science could help people to find a way through to healthy, fruitful living together. By the end of this book, the reader should feel able to join the synergy that New Science describes.

The Age of Reason, as the Enlightenment period is sometimes called, introduced the idea of experimentation *on* the world to gain new knowledge, aiming to show how much of the future is predictable and well ordered. The method relies on controlling or stabilising as many variables as possible in order to observe changes of one specific feature of the world at a time. This way of thinking has vastly increased the human being's power over nature. However, this experimental method, and the type of knowledge it gives us, is powerless to predict what is likely to happen when life gets pushed into turmoil or chaos. Enlightenment Science methods can predict what will happen when water moves smoothly through a pipe at a steady temperature, but it cannot predict what will happen when that water sprays out onto rocks in a turbulent river in strong wind on a rainy day in the depths of winter. The New Science of adaptive dynamic systems, however, *can* make a range of probabilistic predictions. The adaptive systems experimenter then has an improved power of choice over which variable to play with for whichever outcome is desired.

A surprising discovery has been that this predictive capacity, when thinking in New Science terms of many factors interacting, remains remarkably true whether considering how physical systems behave, such as the weather and climate, or biological living systems, and even the social interactions among plants or animals or microbes. This book applies this depth of new understanding to a new level of human interactions with each other. It applies the New Science of adaptive systems to the rough end of emotional literacy, when the dynamics in families, neighbourhoods, organisations, and between nations is getting pushed towards turmoil, chaos, and breakdown, and when the natural environments of the world are being pushed to extremes that are already becoming destructive. How people respond in their hearts and minds, and the actions they take to resolve those tensions or despairs, can be better understood when the principles behind the New Science approach to life are better understood.

In all of these settings, the human power of choice alters life's potentials for healing, renewal, or regeneration. When life turns chaotic, the old Enlightenment Science methods of controlling experimental conditions cannot provide adequate knowledge for useful predictions. Facts about what happens under the tightly controlled conditions chosen by an experimenting team are *not* as generalisable as many believe. In the general mix of life, much reverts to enlightened guesswork in the huge, uncontrolled experiment that life is. Enlightenment Science *controls* experimental situations, dissects life into separable bits, names them to death, then has to work out how to put them back together again. New Science methods keep all the bits together as a dynamic or even living whole, then nudges the system to see how it responds and—vitally—*feeds back* into the observers' lived experience. Attitudes among researchers of life science have moved. There is a paradigm shift, from control attitudes, to relearning wisdom as we influence systems gently *from within*. New Science methods are entirely relational. They observe patterns that change, and predict relational influences.

Enlightenment thinking more generally has greatly enriched life in many respects, in the arts, in rational philosophy, in exploration of the world and now of outer space, and in recording the beauty and diversity of nature. All this has been achieved by increasing the human powers to differentiate one thing from another. But those powers have also created artificial divides. Further problems follow from excessive divisiveness. Divided 'parts thinking' is still taught as 'science' in most schools, as a way to think about material dust and fluids. To influence systems from within, equal weight needs to be given to the question of how the mindful life moves relationally with this dusty water. The *nature of relationality* is addressed in this book, as a core feature of movement, as a core feature of making life-enhancing choices in the midst of changing circumstances. It is only in the last century that the fluidity of energy and matter has been recognised. This fluidity is still not widely appreciated in a practical way by many people when they face difficult decisions. This book aims to shift the focus of attention for decision-making into a New Science paradigm that opens a door into a more creative world.

Habits of thinking learned in schools around the world affect wider life-settings for decades. During the most recent decades many new variables, such as widespread social media connections, have begun to influence life's processes and even sustainability. Relationships break down as much as fizz as a result of remote media misinformation. The New Science of Information and Communications Technology (ICT) has turned the naming of parts into a rapid turnover of ideas instead—memes. Trolls and bots remotely name parts that they don't like in order to control, exclude, and disenfranchise 'them' from the healthy influences that New Science would introduce into life. We are in a transitional phase of humanity's attitudes to knowledge and power. The time is right to gain a clearer picture of a way through to renewable stability in life. This is about the potential within matter for life to re-emerge with new beginnings when the whole has become troubled and shaken.

This book maps out a path that joins or nudges the separated dots of life's experiences into a picture. It is a moving picture that takes shape around the heart-level values that people seldom know how to name because they are not parts. But this empowering capacity to name personal values can help to restore integrating order to life. By 'heart-level' I mean the inner core of a person, the inner dynamic from where personal life experience is generated emotively as our personal values engage with the world. The inner heart, as I shall be calling this concept, is the meeting of our physical, social, and inner ecologies as a whole person.[1] In this book, a way to recognise and name our truly *personal heart-level values* is explained. This method overcomes the divisive tricks that the rational mind can play on our perception of patterns of relationships. Relational patterns vastly influence the choices people make when facing change, but subtly. The method, called Emotional Logic, empowers people to clear a way through the debris of divisive thinking and control attitudes that separate one thing from another, and empowers choices that nudge life towards better integration and adaptability. Equalising power in this way towards mutuality and synergy is not a sign of weakness. It generates life, not deathliness.

***

The more I looked into the subject of life's capacity to heal and renew after divisive setbacks, the deeper and wider my view had to grow. Telling this story of discovery will take you, an enquiring reader, through regions of knowledge that inevitably you will say you know nothing about—yet. I had to explore them too. However, now I can point out more easily how the paths and the dots connect, and speed the journey for others. I cannot say that you will find much affirmation of your current knowledge about the world in the chapters to come. The depths from which personal healing bubbles up after trauma or setbacks have a vastly different nature from the familiar predictable habits of the environments we settle into for much of our lives. The mind plays tricks on us, like covering up our blind spots; but the beauty from which our familiar world emerges is awesome when it is unveiled. When people experience these depths of life in a new way, everything looks different. Then, the renewing possibilities of hope colour life from within more realistically, richly, and honestly.

This is not a science textbook, although it examines science in great depth. It is more of an artistic work, or like a cartographer mapping the personal inscape for the interested participator in life, adding layer upon layer of colour, movement, and texture through which you will see the relevance of New Science to finding creative ways through life's present troubles to joy. While not abandoning the benefits of Enlightenment Science dissection, to integrate the New Science's repatterning of life can balance the future potential for renewal. I use the term 'New Science' to include many colourful areas of exploration that we shall lightly touch on as we gently feel our way forward to find where we belong in this ever-renewing world, such as chaos theory, adaptive systems, emergent order by feedback learning, quantum technologies, information theory, and process philosophy, most of which probably mean very little at the moment but will add to the reader's sense of belonging when explaining how they fit together to empower new beginnings.

On the other hand, there are people who see the science of the last 300 years critically only as a source of ills. For example, some may even go so far as to see a conspiracy in science to undermine ancient wisdoms, and may reject experimental reasoning as if it is of evil intent. My response is that an Enlightenment Science of dissection was a necessary phase before this New Science of integration into life could make sense. The reintegration of different perspectives on life now allows a respectful understanding of spirituality to be rationally considered alongside scientific thought, when the wider heart-level context of life is considered. The approach to understanding movement being presented here can bridge science, as a quest for truth, and a person's spiritual quest for truth. Focus and context are equal partners within the truth of 'living movement'. They dance together. Enlightenment Science has described the dust- and water-based organisation of life in great detail. But this substance, which we now understand has inner movements, is stardust in which the same energy that warms us from the sun's radiance is swirling silently and microscopically round. We too are made of that warmth and stardust. All creatures and microbes of the earth are. But we humans have the greater power to make choices within it all that may not enhance life. And for the moment, we may choose thus mainly for reasons we do not yet understand, but may grow to understand.

***

This is ultimately a pragmatic and practical workbook. It looks at the emotive depths of our life experience to give a robust account of how our human power of choice can change things. Sometimes, knowing which way to turn when all familiarity has been torn away can be difficult. This may become a survival manual.

Our life experience hovers in a social ecology. It is suspended somewhere between the microscopic and the telescopic views of the cosmos that our scientists explore. We co-create our paths through a wilderness by speaking order into a feared potential for chaos, and we cooperate as far as possible to maintain a safe future. Order can emerge from inner depths, however, when we contribute to life's mutual responsiveness. Trust in that process may grow when we are able to see through the veil of rationalised dust in which Enlightenment Science has shrouded human values. Hopefully, by the end of this book that depth of vision will seem natural and spontaneous.

But what of time, and of timing?

My wife Marian and I, in our early thirties before climate concerns overtook international life, went rafting down the Zambesi River's white-water rapids below the Victoria Falls in Zimbabwe, 'the smoke that thunders' as the local population call that amazing wonder. We were carried away by the excitement.

When flying in, I had been fascinated by the deep ravine of the Zambesi. The Batoka Gorge below the Falls is a series of sharp zigzags gradually straightening out. My mind would not settle until I had found out more about its unusual course. I discovered that each zig and zag had been an earlier Victoria Falls. Imagine the timeframes of the erosion needed to cut each rock edge. A new one is being cut now, a sharp angle turning in upstream behind one end of the main Falls.

Here is the point. This reshaping and recession of the waterfall's face slightly further upstream into the flowing Zambesi River is happening because southern Africa as a whole is being lifted slowly upwards by movements of the earth's molten core. This pushes the pressure of water back onto a fault line in the rock—a small fault that grows bigger over millennia. Every zig and zag had been a slowly reshaping wonder.

In those lands, the Saan people, the bushmen, have lived as semi-nomadic hunter-gatherers for somewhere between 20,000 and 100,000 years. It is only in the last 200 years that pressures in the politics of people have forced many Saan to learn an agricultural way of living. New settlers, educated into an Enlightenment Science view that the ecology could be used for economic progress rather than living as part of it, had influenced a change of relatedness. That change has transformed life. The Saan have had to adapt to a change of their prevailing *social environment*, even as the physical environment also is very slowly adapting to larger-scale forces.

In the present days, our third-millennium CE world is going through its own major geopolitical transition. The climate is changing as the world wobbles on its spinning axis. Our growing energy consumption and burgeoning population returns carbon to the atmosphere from long-held stores in the earth. Population migrations and conflict put pressure on others to change, which they resist, sometimes violently. Artificial intelligence and robotics redefine what humanity is. Passivity with fingertip power through social media makes or breaks lives remotely and anonymously.

Choices need to be made. Action needs to be taken. The timing of action may be important.

This book (skip over the bits that on first reading you cannot understand, and gain the overall drift of it instead) explores how at every level of our shared ecology a common set of *principles of organisation* can bring order out of turmoil and chaos. From the pre-microscopic quantum to the cosmological rotation of all that we know, and all that we love, and from the widest sets of social networks that we move among to the inmost depths of our inner hearts where those movements also stir our lives into action or rest, these principles of organisation are at work ceaselessly. The issue is this: do we understand how to cooperate with them, or do we choose to hold on to values that conflict with these principles? Let's find out.

***

In **Part I**, I use the ancient Celtic icon called a *triquetra* to describe the core *principles of organisation* behind the inner workings of the knowing brain and the known world. It explains how the human being emerges to become an orientated conscious person who uniquely perceives their ecology as space, time, and objective reality from their own experiential perspective within it. These principles of organisation are common to both matter and mind. They shape our conversations. Our sense of conversational orientation can easily be disrupted, however, for example by drinking too much alcohol, or by disease or anxiety. The importance of pattern-recognition for organised responsiveness is explained, and the vital importance of feedback loops to develop responsive personal identity and group identity in a changing world. *Synergy* emerges in conversational feedback loops, and in our experiencing of physical movement.

Part II further develops this ecological view of the person. The principles of organisation that have been described can overcome body-mind dualism to integrate them into a mindful walk in the physical and social world. A renewed concept of mind is developed that includes emotion as an equal partner with consciousness, thought, and will. Their integration at a heart level is vital to being a choice-making person. The brain's neocortex contributes analyses of features of life as options for the choice-making process. However, the *values* that shape our decisions when faced with a choice are held deeper, in the limbic and complex core of the brain. This is far, far more advanced in its functioning than the mere reflex reptilian remnant that people are currently mistaught. Values are physically and continuously embedded in the core brain. Personal values are *the priming of filters* to sensory inputs passing through on their way up to the cerebral cortex. They are primed by anticipations from the forebrain's action-planning of what we expect to happen next. Challenges to these personal values emerge from these filters neurologically, when sensory feedback clashes with them, as changing *patterns of loss emotions*. This level of reactivity is preverbal, an inner movement happening below an *orientation horizon*. The brain, like the body's immune system, is a pattern-recognising organ within the wider movements of life. Its various memory systems allow patterns to be compared, which empowers personal choices. Heart-level values and emotions are not problems just to be regulated. Their patterning informs rational decisions by weighting and activating options that match our current values. E-motion is *energy in motion* informed by values. Rather than consciously separating the concepts of energy and information as Enlightenment Science does into physics and information theory, in the inner ecology these integrate as *informenergy,* the impact of personal choice on movement.

In **Part III**, having outlined in Part II the major reshaping of thought needed, I can go on to clarify the mental tricks that hide this heart level of reality. Patterned movements of informenergy connect people at heart level into wider networks of life. We truly are intra-active, choice-making participants. However, the way the cerebral cortex constructs a sense of orientation in space, time, and substantial reality converts our *changing relational experience* into *object recognition*. Naming objects helps conversations generally, but when people mentally take a step back from conversational orientation to analyse life objectively, three different types of dualistic thinking result, which are all logically incompatible with each other. All of these distort personal identity, and create misunderstanding among different group-identities. They overemphasise different types of objectivity. Nevertheless, each type of analysis has its benefits, and when successfully reintegrated into a conversational balance they enrich social and personal life. Becoming habitually stuck in one analytical perspective, however, veils conversational reality, restricts movement, limits values, and disrupts triquetral conversational synergy for the renewal of life.

In **Part IV**, chapters 11 and 12 mirror each other to unite body and mind in a life-enhancing conversational walk. In Chapter 11 the way the physical world has an *implicate order* is described, which may be informationally shaped from within by the core triquetral principles of organisation. Rather than there being any regular 'parts' behind matter, which are imagined to move in relation to each other, a *triquetral process* is described. This first generates synergy as a mutual informational vibration. These vibrations could scale up by harmonic remote entanglement and resonance into the progressive energy waves and particles that we know as movement in the *explicate order* we live in. 3D-printed models of this process have shown how pre-quantum indeterminism at this level of holomovement could scale up into some measure of predictability at an engineering level of physical life. Gravity can be shown interacting with light and heat waves in this informenergy model, iconically representing the fusion reactions happening in radiant stars.

Chapter 12 describes how the same principles of informenergy organisation, when scaled up into our explicate order, can describe the way people influence group dynamics in human social systems. In personal group dynamics the equivalent to fusion in stars involving light, gravity, and heat is the knowable emergence of the *presence of loving kindness* that unites people at a heart level. This is not just the soft, cuddly, cotton-wool view of love, however. Love is understood in Emotional Logic to be enduring relatedness in stative-active modes of joy and grief, grieving actively moving people on separation, brokenness, and misunderstanding to explore how to reconnect. Being the presence of loving kindness is a choice, despite pain, to continue as far as is possible in mutual responsiveness around named, life-enhancing values. This is the natural source of speaking life to each other, by valuing these qualities of character to bridge all human differences and divides, and to allow each other to diversify in a unity of meaningful informenergy. Having that attitude among people who inhabit the land could, in theory, have an informenergy impact on the land to increase its fruitfulness, to thrive even as humanity thrives.

In the **Epilogue** there begins a journey for thoughtful and potentially joyful people to know how and why a compassionate response to others in the next 30 years would physically impact the future course of environmental change more than making a grudging or divisive response. The choice is ours.

Part I

The Knowing Brain and the Known World

Chapter 1

Orientated and Aware

Three sensory association areas of the brain

Your brain has three main *sensory association areas* that feed into your awareness of being yourself. They orientate you, in real time in your lived environment. They are all active while you are reading this. Figure 1 shows where, right now as you read, selected features from all your various senses are mingling with your memories to create a mental background that orientates your *knowing awareness*—where you sit, how your breathing is, the feel of the book or device in your hands.

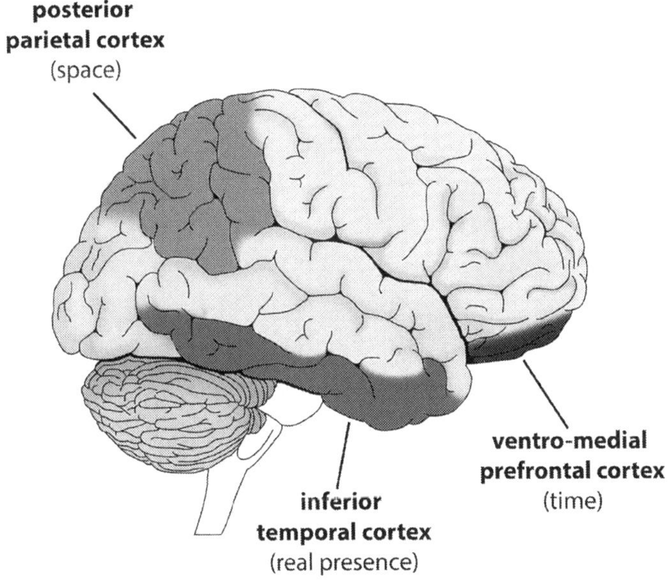

Figure 1: Three sensory association areas of the brain

Your *three sensory association areas* orientate you as a person—in space, in time, and in your experience of reality as the 'presence' of objects known through your senses. Being orientated in space, time, and experiential reality gives your life a context in which to anticipate the changes that may happen next, and with whom. Within this orientating context, you are empowered to make choices about your communications and behaviour in a socially changing, physical world. This book aims to empower you, by depth understanding, to make choices that are life-enhancing.

We know that *being orientated as me in the world* is a product of your brain activity, because consuming alcohol would disrupt that sense of orientation, as it does for other people. People bump into things, lose track of time, and forget things about themselves that they do not want to remember, whenever chemicals such as alcohol or other drugs disrupt the communication between their brain cells. The same can happen after brain injuries, vascular strokes, in dementias, and with some brain infections that alter the brain's structure; also, social orientation is altered with anxiety states or depressions, which arise when the communications chemistry of neural networks gets pushed to extremes of overactivity or depletion.

The astonishing and more remarkable fact than the brain's potential for failure, however, is the daily normality of its success at orientating people enough to communicate with each other sensibly. Out of a potentially chaotic overload of disorganised electrochemical activity in your nerve cell networks emerges an organised impression of *me being me doing what I do in relation to you*. At least, mostly, until confusion or hesitation sets in sometimes. It is quite mind-boggling, if you slow down enough to think about it, that this process of organisation reactivates whenever you wake up from sleep. It is still you stretching and yawning yourself into activity.

The point I shall be making throughout this book is that the *principles of organisation* that orientate your emerging mindfulness of life as you wake up to the reality of the day are embedded structurally in your neural *networks*.[1] And because other people's brains have a similar physical structure, sensible communication can occur between people who see the world differently in detail, but whose thoughts nevertheless become organised in space, time, and experiential reality according to identical principles.

Throughout this book I shall be using the ancient Celtic icon called a triquetra to display these core *principles of organisation* (Figure 2). By adding different labels to it I can explain how they appear in different settings and shape the movements there. Tri- = three; quetra = corners; a triquetra is a three-cornered shape that is not a triangle (and therefore it is not a triquetrum). The plural is triquetrae. A triquetra is an icon of the ceaseless dynamic that processes within all that seems constant in life. Its three corners and loops show endless movement.

Figure 2: The triquetra—an icon of dynamics

Figure 2 is colour-coded to match Figure 1, which shows as different tones of greyscale in the diagrams printed here. (The colour diagrams can be downloaded as a PDF from the link shown in the List of Figures.) The three sensory association areas are coloured orange, purple, and green. This is to show the theory and hypothesis that core triquetral principles of organisation (Figure 2) are sufficient (and necessary) to explain the extraordinary fact that stable conscious orientation can emerge from potentially chaotic brain electrochemistry when structured in this way. In each sensory association area, simply pairing two out of the three core features of sensory and memorised life could construct the three different types of contribution to conscious orientation, as we shall go on to see. The importance of this is that, while this adaptive system can disintegrate in various ways to create illnesses and other problems for humanity, the simplicity of the system also makes it robust enough to recover and heal when given the opportunity.

These same colours (or matching greyscales) are seen in the triquetra. Two of the 'corners' associate in the loop shown, connecting them as a pair. The loops represent the integrating process ongoing in each corresponding sensory association area to construct the consistent impression of being a person orientated spatially and relationally in a changing ecology. No energy is wasted in this simplest of complex informational processes.

Informational patterning in the triquetra

The three orientating features for orientation extracted from the 'distance senses' of vision, hearing, and skin sensations are *changes* (on-off), *relatedness* (synchronicity), and patterned *forms* (edges, corners, harmonics). Primary sensory reception areas in the cerebral cortex identify these and other features within each sense modality. In their neighbouring secondary sensory reception areas, these are grouped and categorised into edges, harmonics, and so on, then forwarded into the complex neural feedback loops shown in the triquetra that distribute them into the sensory association areas shown in Figure 1. Here, mixed sensory processing starts to integrate them. They will mix also with features extracted from the inner 'proprioceptive senses' (muscle tension, joint position, joint movement, balance and position senses) and with reactivated memories, so that distance awareness becomes associated with inner awareness.

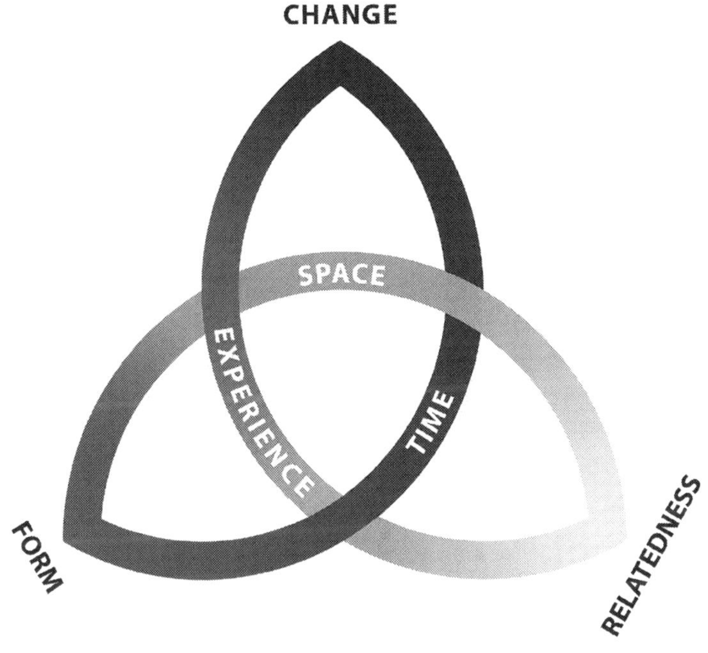

Figure 3: Three primary constructs for orientation

In Figure 3, change, relatedness, and form are represented informationally at the three corners of the triquetra as inputs to this orientating system. The three semicircular loops extending from them display the pairing of inputs in slightly different patterns in the three sensory association areas that are structurally coherent with the triquetral principles of organisation. If this theory is correct, this foundational system would be stable enough to have a profound impact on mindfulness, potentially regenerating or renewing the experiential notions of space, time, and object reality even after a crisis of some sort may have caused a mental breakdown and temporary disorientation.

The three primary colours are used consistently throughout this theory to represent the inputs to *any* system of *change* at every scale of life and physics (blue; dark in greyscale), *relatedness* (yellow; pale in greyscale), and *form* (red; mid-tone in greyscale). Artists can mix the full spectral range of colours in a colour circle from the pigments that absorb and then transmit light consistently as these three primary colours.[2]

In the parietal lobe sensory association area, the pairing of relatedness (yellow) and form (red) information can generate an impression of space, shown as the triquetral loop coloured orange in its mid-part. People naturally use spatial language to describe the relationship of forms, whether in the physical ecology (depth, height, width, length), the social dynamics of life (hierarchies of people's roles and relationships), or in the inner mental relationship of ideas to each other (the logical structures that philosophers employ).

In the infero-medial prefrontal lobe sensory association area, the pairing of change (blue) and form (red) information takes place, shown as the triquetral loop coloured purple in its mid-part. Changes of patterned forms may have an infinite diversity, but many such changes are cyclical or repetitively vibrational, such as a pendulum swinging, or the sun rising and setting, or the pulse felt at your wrist. Changes may also be single episodes, or progressive developments, or growth. Comparing one patterned form with its *transformed pattern* creates an impression of time's passage. This awareness of potential for change of forms enables choices to be made about the timing of action to achieve an anticipated outcome, as movements progress or pass through their cycles. The relational interaction between two cyclically changing forms may produce harmonisation and resonant magnification if the phases cohere, or negation if they are out of phase. The sequential timing of a movement alters the course of competitive sports such as tennis, for example. As events transform over time, their relatedness is knowable *in time*.

In the temporal lobe sensory association area, the pairing of change (blue) and relatedness (yellow) information generates the experience of *object recognition*, shown as the triquetral loop coloured green in its mid-part. Epileptic states affecting this part of the brain can produce a sense of unreality in lived experience, in which phenomena remain merely as changes that seem to bear no relation to each other or to past experience, a mental state known as depersonalisation. Object recognition is an informational process in which connections (relatedness) are made between, for example, the patterned sensory form of the way a bicycle looks from one angle with a vastly different sensory pattern when it is viewed relationally from different angles, for example from our two eyes simultaneously. The way the *changes relate to each other* creates the awareness of relational consistency that *is* a patterned object. Another example is that to feel the quality of the material cloth of a garment we do not merely push a fingertip against it. We move a hand across it, moving through it, so to speak, presenting the brain with a relationally changing range of sensory data that becomes organised into and memorised as the objective cloth curling around our hands. The words *experiential* and *experience* have a source meaning that is important to understand. In English these words derive from the Latin *ex-* meaning 'from', and *pereo*, meaning 'I go through'. The word *experience* means 'generated from going through a situation'. A sense of objectivity is generated by going through situations experientially.

This third orientating feature of lived experience (creating an objective reality out of changing qualities of relatedness) is the great strength of consciousness *above an orientation horizon*. It clarifies anticipated options of behaviour to inform conversations and create a power of choice between planned actions. However, we shall be going on to explore in great depth how *below this orientation horizon* this capacity of the brain to objectify features of our wider relatedness can become also our greatest human weakness. It is heavily implicated in the mental tricks that hide reality from our heart-level awareness. We shall explore awareness in Part II, the mental tricks that veil it in Part III, and find ways to overcome this weak spot in Part IV.

Physical, social, and inner ecologies

To prepare a way to the solutions described in Part IV, I need to explain how these core principles of organisation can be seen not only in the way the brain could generate that sense of orientation out of changing sensory data, but also in the way those same triquetral informational processes could shape all physical processes in our material environments. This is an ecological view of the human person, and indeed of all life-forms, the human person having the greatest range of potential choice for adaptability. I am, no less, proposing that identical principles of organisation apply to both mind and matter.

People's chosen behaviour can engage effectively with the physical and social world *because* these principles cohere, I propose. If so, then this speculation would also allow the possibility of times of incoherence between mind and matter. People can drift off into daydreams rather than remaining mindfully engaged with reality. In fact, as we shall go on to see in Part III, there are numerous ways that this conversational engagement system between orientated person and environmental ecology might drift off into unbalanced ways of trying to make sense of life's events, even though it can rapidly restore stability and wakefulness when an unexpected change of relatedness requires attention.

I am proposing that triquetral principles of organisation, when understood in greater depth, can be seen to guide processes at every level of the material and energetic world, from the subatomic to the universal. Until this hypothesis is disproven, we can assume for the duration of this book that your knowing brain, and the known physical world as you choose your way through it with all its living social complexity, are equally organising energy and information *in the same patterning way*.

It will be helpful early on for the theme of this book to expand the meaning of 'patterning energy and information' into three ecologies, which together integrate into one overall ecology of life. This can help to identify which bits of life's ecology are orientated by the brain into a *space, time, and experiential presence* mental framework, and which features of the ecology of life are not orientated in that way. I shall be introducing a new term shortly, the *orientation horizon*, which is intended to help people to categorise their own experience into that which is located in space and time, and that which is not.

Firstly, there is the world's material and energy ecology. Movements of this physical ecology preceded our birth, will outlive our death, and process now beyond our awareness with regular and deterministic changing patterns that are described as the Laws of Physics or Nature. The statements of these Laws, however, are conversational agreements made between mindful human beings when comparing their lived experiences and observations in different times and places. The 'Laws' change subtly as the scientific conversational context expands with increasing technology. This process of comparison has lasted over centuries or even millennia of careful observation of the sun, moon, and stars, and measurement comparisons using a common language called mathematics. Mathematics is the language of physics, and it evolves in time just as any shared language evolves. The important point is that this *physical ecology* transcends the individual, and measurement of it is a social agreement.

Secondly, there is the social ecology, which includes not only other human beings, but also all the infinitely diversifying changing forms of life that emerge as material-energy patterns for a time within the world's physical ecology. In their own timeframe, all living forms return back into its dust eventually. This is the ecological view of all living organisms within the wider physical ecology, whether a creature, plant, or bacterium, and applying equally to the embodied human being. The principles of organisation in this social ecology are identifiable in the *moving patterns of communications* needed for life-forms to emerge as networked energy and information within shared physical environments. Organisms compete with each other for the environmental resources needed for each moving form to thrive. The individuals of some species also learn to cooperate with other living creatures to manage those physical environments. They learn how to feed the physical ecology that in time will feed more of the resources that they seek back to them. At the human life level this process might be called cultivating the garden, but it can corrupt into abusing the ecology if the principles of organisation become distorted and unbalanced. The social ecology is constantly repatterning with communications needed to hide from predators or to cooperate into societies. Triquetral organisation of movement creates *systems of communication* among multilayered living forms. The human being could not survive without healthy colonies of bacteria in their guts and on their skin. Communications within the social ecology may be chemical, electrophysical, verbal, obvious, or subtly silent. Movement, as a philosophical concept, may be consciously known by *changing relatedness of patterned forms*. The social ecology is an organisation of changing patterns of relatedness by feedback communications. This ecology may become consciously known by a person to improve the quality of individual choices.

Thirdly, there is the inner ecology of an individual life-form. The cellular structure of living tissues is a profoundly intense and active dynamic of biochemistry and biophysics in constant turnover. Hundreds, no thousands, of chemical reactions and physical wave-field coherences self-organise at every level of order you care to imagine into a unity within a living cell, even within that of a single amoeba. Scaled up, these communication feedback patterns are even more astonishing as they organise into an embodied human being according to the same principles that apply across the whole spectrum of emergent life. There are about 100 billion neurons in the average human brain cooperating and coordinating their physical chemistry. We shall consider how triquetral patterns may organise their feedback communications, and how this self-organising brain fits within its *embodied* human being, who is a continuous gathering of dust and water into about 37 trillion living cells. This mindful body needs to coordinate to survive and thrive in a social ecology. Effective feedback loops of information are vital for this.

The internal ecology has its own patterns of movement, however, which emerge within the shared social and physical ecologies as observable behaviour and relational harmonies or disharmonies. Each individual human being has choice over how to communicate externally their internal balance and their memories. People tell the story of their experiences in words, actions, and subtle silent communications within their wider feedback networks. The complexity is awesome. Its integration and coherence with others and with physical nature is potentially the beauty within any feature of life that you might care to focus your attention upon. Sadly, that beauty can also be marred by tragedy or trauma, by distorted character, and by misunderstanding. However, life within does have a resilience to adapt and renew after setbacks. Healing and restoration of the inner ecology and social network are possible *because* those core principles of organisation are foundational for both matter and life's re-emergence after tragedy. It is as if they are an inner light that also transcends the individual and cannot be extinguished.

Alerting, searching, anticipating—the orientation horizon

The dynamics within these coherent ecologies are all active, to some variable extent, even while you as a person are asleep. They shift into a different mode on waking. On waking, or becoming more alert, people gradually (or perhaps suddenly) restore the impression of being orientated in a consistent environment and gradually organise their thoughts and feelings and physical appearance into being a coherent, consistent person ready to go out and meet others. Here we revisit the notion of an orientation horizon. Some experiential events in life become rapidly located within a mental framework of space and time, and others curiously do not.

Memories, imaginations, daydreams, speculations, planning, and intuitions, for example, may be informationally processed through other parts of the brain than the three sensory association areas outlined in Figure 1. The sense of smell, for example, is not localised because it is processed elsewhere, but in such a way that emotive memories may activate an orientating search to locate its source in order to move towards or away from it. Orientating a search for the source of an aroma does not remain in the sensory neocortex. The three sensory association areas all project further forward into the frontal lobes. Here they integrate with each other *around the planning of a bodily 'motor response'*.

This initiation of *movement* is represented by the central area of the triquetra. Here, the inner ecology is continuous deeply with the social and physical ecologies. There is going to be an informational transfer between them that will reshape the patterning of both the physical and the social ecologies. A sense of smell may alert a search that self-organises a mental framework with sufficient orientation to identify, for example, that food is overcooking in the kitchen, and perhaps starting to burn. Verbal communications will probably follow, probably not very subtly. A spreading wave of repatterning activity through the relevant social and physical ecologies will follow. All of this process involves feedback loops at every step of the way.

Mostly, the human person becomes orientated in order to physically engage in conversational movements or exchanges in the social ecology. Conversational orientation exchanges the inner story about an infinite range of experiences, not only with other humans, but also with pet or other animals, who may coherently benefit informationally from this 'social feeding' and feedback. Some gardeners talk to their plants, subtly or openly, hoping to aid their growth and beauty. Many people talk to 'the universe', however they imagine their widest possible physical and social ecology to be. Some will perceive spiritual personalities, gods, or a God in their universal Big Picture of the social and physical ecologies. The domain of spirituality does not need to be other than these three ecologies (physical, social, and inner) when the vital feature of feedback communication is considered to be connecting all three of them *below the orientation horizon*. For the duration of this book, I shall equate spirituality with the relatedness feature of an individual's Big Picture of life. This inclusive, ecological concept of spirituality enables me to describe even humanism and atheism as the qualities of an individual's spirituality. Because of the heart-level integration of relatedness with transformation (change of patterned forms of living) in a concept of movement, a person's spirituality may be thought of as a quality of movement in their inner heart.

The key issue here is the activation of a search on becoming alert. Life is exploratory, not merely passive. The alerting feature of life is not in itself orientated. Wakeful awareness as an individual can lead on to becoming conversationally orientated and alert as a person-moving-in-an-ecology. This is emergence above the orientation horizon from heart-level qualities below it.

The fact of this experiential transition leads me to make a distinction between *awareness* as an individual experience that may not be orientated, and *consciousness* as a shared experience in an ecology that probably is orientated to some extent. The Latin *con-* (with) and *scire* (to know) suggests responsive movement *with another*, whether that is in the physical or the social ecologies. Relatedness in movement is a vital feature of consciousness. This is not a rigid separation of concepts, but a functional description to aid noticing where one's own mental framework is shaping events with an orientation horizon.

If you reflect for a moment on your own experience of identifying this orientation horizon, you may notice that in the very moment of reflection you will have disengaged your neural orientation mechanism. The behavioural consequence will probably have been a reduction of movement. Even a pause for thought may disengage orientation. You may prefer to give the pre-orientated state of living a different descriptive name to being *under the orientation horizon*. Some personalities may prefer such terms as 'having higher thoughts', or 'ascending into the realm of pure reason'. However, for the purpose of making life-enhancing choices when facing difficult situations, there are good reasons for identifying a horizon below which we need to slip in order to create a gap in time for reflective thought. The theme throughout this book is how to fill the gap in time needed to balance out life-enhancing choices with an awareness of named personal values.

The core limbic brain

I shall employ Paul MacLean's categorisation of the brain into three main functional parts to illuminate relevant features of the way the inner and social ecologies communicate.[3] It is a 'broad-brush' description for a general readership of how the brain and the social messaging systems of the living body move together. It cannot be bettered as a helpful starting place to understand the brain's *connectivity*. Specialists with a detailed knowledge of neurobiology will take issue with its details, but the broad categories are widely accepted (but still misrepresented in some settings such as schools). Further details about brain structure will be added when they become relevant.

In particular, professional neuroscientists no longer accept MacLean's original use of this model to describe the *evolution* of the brain. This is an important point. Popular teaching even among psychologists still talks of the reptilian brain, which is a very unhelpful concept and easily misunderstood. It distorts the picture of how personal choices are made. The potential impact of this misunderstanding is so profound that some neuroscientists who still use this term claim that there is no such thing as free choice or free will. We shall return to sink that idea later.

In Figure 1, the stalk at the bottom of the image is the *brainstem*. Here the heart rate and breathing rate are regulated, along with various other body functions necessary for survival of the organism in its physical ecology. This is the part that some psychologists still unhelpfully call the reptilian brain, because its functions are mostly reflex. Unfortunately, many teachers of popular science extend that idea much further up in their mental picture of what is going on inside the head. Another very important feature of the brainstem is the *ascending reticular activating system* (ARAS), which is the alerting system we have just been considering. The ARAS sends fibres from the brainstem directly up to the top surface of the cerebral cortex to prepare the cortex to process new inputs. Most other pathways have synaptic junctions on the way up through the midbrain,[4] which slows their path, but adds options to modify the ascending sensory data, introducing intelligence over reflex.

The second functional region of the brain to consider is the *neocortex*, the extensively folded dome of tissue over the top, with its left and right cerebral hemispheres. The surface 3–4mm 'cortex' of this is the 'grey matter', the dense and highly complex neural network through which waves of repatterning electrochemical activity can spread informationally across the brain's surface. The three sensory association areas highlighted in Figure 1 are part of this rich, feedback neural network. Here in the cerebral cortex, the detailed analysis, categorising, and imaginative planning with memory recall is part of the inner ecology that diversifies life.

The third functional part of the brain in MacLean's model I shall call the *core brain*. In Figure 1 this is partially visible as the two thickened areas at the base of the neocortex. These hint at the presence of an extensive and extraordinarily complex subcortical network of structures and connections with a multitude of roles that we do not need to detail here. Apart from those concerned with smoothing movement and integrating it with sensory inputs, I shall be referring to its core brain's limbic system, shown in Figure 4.

The limbic system is active during emotional processing and memory, so it is seen to be active almost always during brain scanning. Most significantly, its hypothalamus translates the overall balance of brain activity into whole-body physiology by regulating the release of hormones into the bloodstream from the nearby pituitary gland. Hormone release is very largely driven by emotion rather than reason, from below the orientation horizon rather than above it.

Orientated to prepare for action or rest

The main point of using this simplified model is this. The core brain integrates *social connection* by balancing sensory input, memory, and behaviour output. Emotions happen in this balancing act.

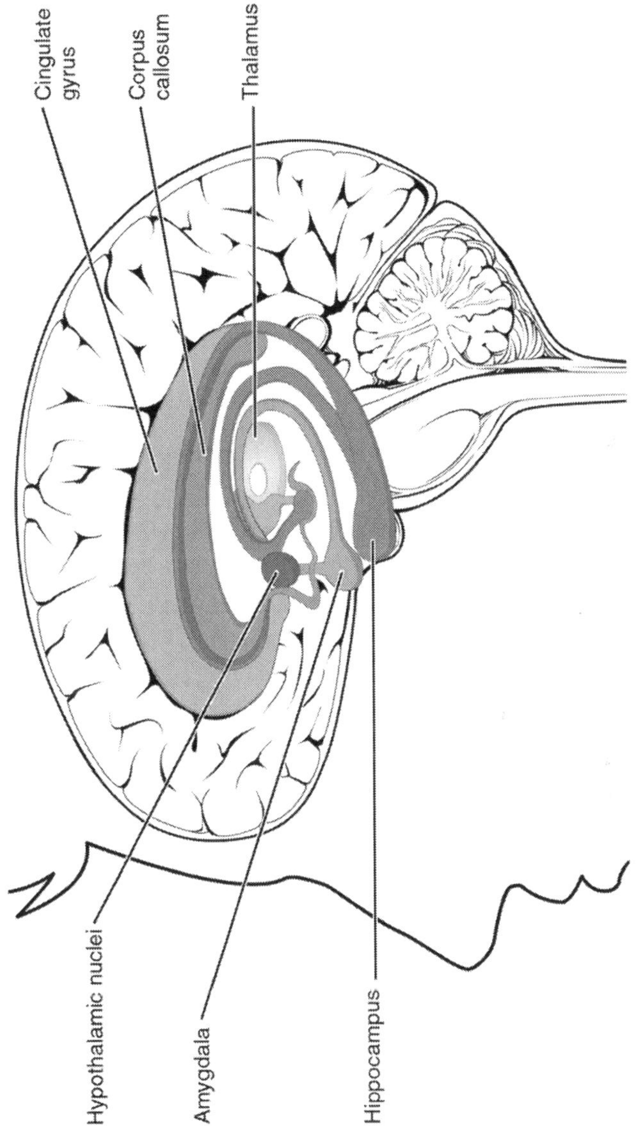

Figure 4: The core brain's limbic system

Social messaging between humans mostly occurs by emotive movement, transmitted physiologically from the core limbic brain to other sentient creatures. This social messaging is a feature of emotion forgotten by those who focus their attention cognitively on only their internal 'feelings of emotion', a distinction that is to be explained more in Part II. Why does an emotion have a physical facial expression, and postural body language that can be read by others, and variations of tone of voice (paraverbal language) that carry more meaning than the words used, and a release of chemicals in perspiration called pheromones that informationally convey basic physical emotional preparations for action to other sentient creatures? It is because the primary purpose of emotion is to alert and relationally knit sentient creatures together with the responsiveness and adaptability to survive and thrive *in the social ecology*.

The core brain is literally the nerve centre where the inner ecology and the social ecology meet, or collide, especially when things are changing fast. That collision in the core brain is emotive, activating, awakening. Reasoning in the neocortex is activated *secondarily to* core brain social balancing and communications. People become cortically aware or conscious *that they are having an emotional reaction*. They then have to rouse and orientate themselves sometimes to try to work out why, or sink into a self-absorbed experiential state that disengages from conversational exchange in the social and physical ecologies. Some neuroscientists even call the core brain the *social brain*. It is definitely NOT reptilian. In advanced societies, the core social brain has a developed *unifying* role that integrates a sense of personal identity in relation to that society, as we shall see.

Behavioural output balancing in the core brain

The three sensory association areas shown in Figure 1 do not, surprisingly, directly connect with each other to bring about a fully integrated sense of personal identity. An important feature of neuropsychology is that they project forwards, and integrate in the frontal lobes' 'motor planning' areas. In other words, a person becomes integrated around the actions they are planning.

Every part of the frontal lobe motor-planning area also sends fibres back to the three sensory association areas, in a bootlacing way, so that reciprocal feedback loops connect them informationally both ways. Action planning is thus continuously reframed in relation to changing sensory experience. Unfortunately, this can mean that people live in a mental world where life is the way they imagine it to be, unless social feedback contradicts this experience. A conversation activates this whole inner complex system embedded informationally in both the front third and the rear two-thirds of the brain. However, this system does connect with the wider physical and social ecologies in a way that philosophers may not have seen.

Conversations, and many physical activities, require anticipation in order to target planned action. This is where the core limbic brain has a surprisingly central role that we shall be referring to throughout this book on making life-enhancing choices. The three sensory association areas integrate not only by projecting forward to the frontal cortex for motor planning, but also they project down into the core of the brain to integrate in a myriad of 'ganglia' (and via them to the cerebellum), to prime body posture. This provides a stable, orientated, physical base from which a person can put in process the detailed actions being planned in their frontal lobes.

Creating this stable base is a far more important triquetral process of informational movement than many people give credit for. It is fine-tuned in many spiritual, artistic, and martial arts practices.

But now comes the main point of this section.

The frontal motor cortex outputs that move the musculoskeletal body are patterned to achieve an anticipated effect. The descending pathways for this also send out side-shoots, which connect with the ascending sensory pathways passing up through the spinal cord and core brain. They land on the synaptic junctions between sensory neurons on their way up to the neocortex. They thus modulate the activity of those synaptic connections, adding higher-level intelligent modulations to local sensory input and reflexes. This means that the core brain's sensory synaptic junctions act as filters on the incoming raw sensory data. These filters are primed by the outgoing motor anticipation. A focus of motor anticipation, imagination, is set within its context of a stable postural base of the body. There is a dance between body posture and imagination.

The sensory filters therefore hold an *anticipation pattern* for behaviour in a dance between focus and context. This is where the inner ecology informationally meets the incoming social and physical ecologies. Other people are on the dance floor too. Unexpected incoming sensory truth, about movements of the wider social and physical ecologies below the orientation horizon, collides in the core limbic brain with a person's hope.

Sensory input and memory balancing in the core brain

Loss of consciousness results from pressure on, or rotational trauma to, the core brain, not injury to the neocortex, which causes only small deficits in responsivity. The core brain's integrating and balancing roles are central to the filtering of conscious awareness.

Sensory input pathways from peripheral sense organs up to the primary cerebral cortex sensory areas pass through the core brain's thalamus, and radiate up to the cortex like a spreading fan of long nerve fibres. Every area of the cortex sends reverse direction fibres back to the thalamus (thalamo-cortical feedback loops), but also to other areas of the highly complex core limbic brain. I shall call these other feedback loops the *cortico-limbic feedback loops.*[5] Within the neocortex there are also innumerable other *cortico-cortical feedback loops*. They functionally hold its parts together in continuously mutual *informational movement,* like a fabric that may seem to have no substance, although it does have substance in the biophysics and biochemistry of the responsive neural networks.

Your three sensory association areas of the cortex therefore have a key role via their cortico-limbic feedback loops to add orientating intelligence to the emotive social messaging process. Resonance with memories occurs by similar projection down from the cortex into parts of the core brain such as the hippocampus. Memorised emotional responses in similar previous situations stored here contribute to the awareness of yourself as a constant person in a changing world, perhaps inappropriately for the present circumstances.

Within the orientating mental framework, a person's wider sensorium of smells, tastes, inner sensations, music, colours, and so on, all add richness to the orientated experience. The *mismatches* between this rich sensory experience and the formerly established *anticipation pattern* become magnified by the sensory filters in your core brain relay stations, where anticipation patterns are held. Those magnified mismatches energise an emotive response within the core brain, while synchronously relaying up to the neocortex. On arrival at the neocortex, the details can be analysed and focused in ways that may subsequently modulate the core brain's response.

Personal unity, personal identity, develops *where muscular movement is primed and planned in socially emotive settings*. This is an important point. We become orientated as an integrated person and aware in space, time, and experience by *preparing for action* responsively with others in our shared physical world. We may choose not to follow the prepared action through, so that action remains only in imagination, or in the replay of memories with slight variants.

Character develops (personal formation) in the inner ecology to modulate innate personality (responsivity to wider relatedness) as we prepare for action (change) in an ecological context. The inner formation of personal identity is shaped by triquetral principles of organisation active within the three cortical sensory association areas projecting (a) forward into the frontal lobes *and* (b) down into the core limbic brain simultaneously, so that potentially chaotic sensory inputs become unified into a sense of 'me' being me as an active agent in a personal and material world.

Feedback loops create patterns of connection

Modern brain-scanning techniques reveal in remarkable detail how real-time neural patterns change as people think or feel or will or choose. Unfortunately, because the core brain is active all the time, it tends not to attract the interest of researchers who are trying to answer very specific research questions. Attention goes to the neocortex instead, to where a tendency for localised activity can be identified when a specific function is repeatedly stimulated.

This leads to a sad conclusion in many people's objectifying minds. By observing that in, say, 80% of people studied, a particular area 'lights up' 80% of the time, it is too easy to conclude that the function studied is 'located' in that part of the neocortex. However, any locality in the neocortex is just the top end of the changing relatedness of patterned neural activity running through multiple feedback loops. Each locality has its own 'relational frame' of feedback loops, within which electrochemical informational movements unite the physical and internal (mental) ecologies. In Part IV we shall see how, by recognising the triquetral organisation of movement deeper than biochemistry, in informational quantum changes within the biochemistry of neural activity, the separated dualism of mind and matter becomes unnecessary, and even unhelpful.

Specific functions do not take place *in* specific locations of the brain; they tend to run *through* specific locations as they extend their patterned qualities of relatedness. The triquetra can represent any of these 'movements'. *Neuroplasticity* means that many functions that seem to be located in particular areas of cortex can transfer elsewhere to some extent with learning. The variability of location is as important as the averaging of location. Areas of the neocortex that in most people tend to be used for a particular purpose can be harnessed to process other functions. This is one of many features of the learning capacity of the brain. Brains are as adaptable as the people who communicate with them.

Mindful engagement of an individual's inner ecology with their changing physical and social environments involves *patterns* of organisation that *transform* by feedback between ecologies. In the inner ecology, a person's ideas and thoughts are informationally changing patterns of relational connections in the brain between various centres of activity. The changing patterns include feedback loop connections of ongoing sensations to memory traces and anticipation patterns, all of which are distributed through the brain's many networks. Therefore, ideas, thoughts, and feelings are informationally embedded as an inner ecology of networked movements triquetrally organising the body's own physical ecology.

For millennia, philosophers have struggled to break through this dilemma, unable by rational thought alone to demonstrate a genuine connection between mind and matter. The New Science since the 1970s, however, when combined with new technologies of brain scanning and applied to living dynamics, has empowered a new way to understand the bi-directionality of informational flow that mindfully connects human beings into their lived contexts. The inner human heart has been excluded from serious consideration for centuries by rationalist philosophers, who prefer instead to focus on the details of cognition, on theories of knowledge, and on logical linear reasoning and behavioural self-control rather than on the relationally patterned transformations that will promote life below the orientation horizon. The New Science, of emergent patterned order by feedback learning, now offers a more heart-strengthening future.

Social feedback loops link world with brain

I was once accepted onto a PhD programme for cancer research to study the immunology of cell membranes, and how communication between populations of cells can break down at the points where cells interface. Having started to understand how cellular dis-connection contributed to disease, I realised that the same truth applied to whole people. Preferring to be out in the social mix rather than in a laboratory for the next few decades, I changed professional direction and became a research-minded medic. I explored how communication patterns and responsive feedback can break down in families and between human populations and between neighbours, and thus harm their physical and mental health. The same principles of organisation apply in both settings at different scales. Social feedback loops of communication between people can generate a hope for wholeness and an *awareness* of brokenness.

May I introduce a cameo of a scenario that actually happened? A man was mystified that his neighbour, whom he did not know well, had started casting hateful glances in his direction whenever they happened to see each other across the boundary between their properties. After a day or two of this the man took courage and said across the fence to him that he had noticed his neighbour seemed upset about something. The neighbour replied that the way he had started singing and smiling was irritating him, and could he please stop it. This stunning response led to a few moments of silence while he gathered his thoughts, and reorientated. His response as he emerged from that shocked gap in time was life-enhancing. He said, "What has happened to you? You are reacting as if something important to you has been lost." The outcome was deeply emotional. That neighbour's wife had been admitted to hospital with a recurrence of cancer. The happier man was able to explain that his daughter had become engaged to be married, but now he wanted to be a friend to his neighbour also. Could they meet and share a small meal to talk further?

The point of this cameo is to picture the feedback cycle of a social engagement in the wider physical ecology. The man had not recognised how his social messages were being received. Other people, and some animals, produce unpredictable responses to *our* preparations for movement. To detect *their* responses to *our* preparations for movement requires our sensory sensitivity, and curiosity about mismatches within that sensorium from our hopes or anticipations. This is the other type of feedback loop that enters the limbic brain, the *social feedback loop* that collides with the inner ecology's *cortico-limbic feedback loop* in the core brain's sensory filters where its anticipation pattern is embedded.

Cortico-limbic feedback loops prepare our outwardly moving behaviour. Enlightenment psychologists, and positivist philosophers of that era whose interest was to understand the actions of a material individual, limited their curiosity at the objective boundary of the body's musculature movements and behavioural postures. They regained interest only at how the sensory receptors in the skin and retina worked, studying how stimuli there activated the neocortex. (In those days of the 1970s, hearing and balance was far too complex to make sense of!) Behaviourism had introduced a simple stimulus-response, linear view of emotional processing, in which both the brain and the social ecology are like mysterious black boxes, the inner working of which could not be controlled enough for meaningful Enlightenment Science experiments. The restrictive view that emotion is a side effect of thought is a by-product of this linear, behaviourist thinking.

However, now New Science neuroscientists and process philosophers understand adaptive system communication concepts. They have more advanced real-time measuring tools to make sense of what happens both in the brain and in the wider physical and social ecologies in response to our behaviour. Inside both the black boxes, *patterning transformations* of *systemically relational life* can now be mapped. They can also be nudged towards life-enhancing responses.

The 'core social brain' thus integrates *two sets of feedback loops*—the cortico-limbic feedback loops that embed the inner ecology as an evolving anticipation pattern, and the social ecology feedback loops that reveal how reality mismatches our imaginations. Each set of feedback loops inputs different informational processes. These collide in the core brain and limbic systems and there generate a sense of being an emotively responsive person embedded in a physical and social world.

Unity in the limbic brain; diversity in the neocortex

It is sometimes hard to grasp that the people we rub shoulders with may each be forming hugely different spatial impressions of the setting in which we live. They may be considering different timeframes in which to plan. They may be aware of different realities in any situation we all may face together, or choose to deny features of them. This diversity is powerful for our survival when in groups that are socially communicating and able to learn from each other. It leads to greater adaptability in changing life circumstances.

The higher brain (the left neocortex in particular) is actively differentiating one thing from another and fixing those divergences with words. The fact that the meaning of words drifts or varies over time among groups of people indicates that words are symbolic labels assigned with a purpose to separate objects one from another. Word labels (symbols) empower ways to manipulate objects. The right hemisphere does not use language to separate. It categorises and groups ideas instead, according to qualities and functional potentials. So the left brain would name some structure as a chair and another as a table, while the right brain will be functionally categorising the chair as something to sit on, and the table as something to put things on. Of course, a person may choose to sit on the table and put things on the chair, which would stimulate learning and recategorisation in the right neocortex to appreciate a further feature of diversity. However, please note that the left neocortex naming of the recognised object as table and chair will probably remain unchanged. The right hemisphere will have been more adaptable to expand its preverbal 'relational frame' for those labels. The name (unchanged) of the object will now carry a different range of qualities and functional potentials through into the feedback loops of its neural networks, harmonising and resonating with different memories.

Diversity of perspectives on life in groups, where people have different relational frames for the same words, is not a problem if there is a conversational reality of openness to hearing others and listening to them sequentially taking turns. However, there *is* a problem if people do not first recognise a heart-level depth of unity below their orientation horizons, leading to curiosity. That preverbal level of social unity moves as a potential for group morale. There is an intuitive level to it. The absence of morale is knowable, even if it is an intangible feature of process. It deserves our attention to explore and remove hindrances. Without intuiting the potential for heart-level unity in life settings, people will not prioritise communicating at a reasonable depth below the orientation horizon. Healthy curiosity, however, can initiate a process of exploring where synergy emerges, from within the relatedness below the orientation horizon, and contributes to *renewing life*.

Chapter 2

An Informational View of Personhood

Information is a verb misused as a noun

A curious feature of human brains at work is that, out of a continuous hum of dynamically electrochemical neural activity and rapid changes of blood flow to oxygenate and sugar-feed it, brains create an organised pictorial mind map of *things*, objects that may seem to be unchanging. Brains can make nouns out of movements. They can make fixed ideas, while nature has already moved on.

Science has shown that *things* in nature are constantly changing at quantum, atomic, and molecular levels, despite the mental tendency for 'object recognition' to create ideas that things may seem unchanging. The converse is also true, however, that the healthy parallel processing of brain and world can dislocate in the opposite direction. People can become 'lost in thought', to such an extent that thoughts may even race away in anxiety states to make a disassociated mental world of impressions while, at the nature scale of physical human life, the physical environment that feeds this thinking brain has truthfully not changed anything like as much! "Come back to your senses" can be good advice in such circumstances to restore mental order out of chaos. In the core limbic brain, it is the mutually responsive interaction between the two principal feedback loops just described (cortico-limbic and social feedback) that enables constructive order to emerge from brains.

But human brains insist on making nouns out of verbs.[1] Within sometimes rapid changes of relationships, the brain detects patterns of activity. It can memorise and name a formed structure in the wider context of life that may indeed endure, or recur, through times of change. The brain is a pattern-recognising organ. That is how it integrates and organises learned responses. People may recognise this tendency in their desire to anticipate and predict outcomes. There tends to be a static mental picture of a preferred outcome. This patterned mental imagination is commonly associated with the recall of an emotive state, perhaps to enhance a sense of security or pleasure, for example, or perhaps a rather scary opposite.

The idea of stillness conceals a stative, patterned potential for changing relationships to emerge. For example, when people sit down they create a lap where their upper legs are horizontal. The lap (a noun) is a place of stillness, but when the person stands up, their lap disappears. The noun of a *lap* is a temporary relational construct. Brains create nouns out of uses, potentials, and actions, defining their boundaries and their qualities by the type of responses that we make *in relation to the felt dynamics of the thing*. This feature of brain function has been termed 'relational frame theory'.[2] Laps are for putting other things on, or for children to sit on. Without those functions, your lap might as well not exist. Indeed, in English, people would not call it a lap when they are sitting down if they had no intention to consider putting anything on it. And indeed, it does not 'exist' until, for example, a growing child wants to show you a toy they are playing with. Then the lap relationally appears in space for a time.

The wide range of nouns used by adults appear to have all sorts of intrinsic qualities that have been subtly remembered from the wide range of exploratory toddler experiences, such as its feel when held in a hand or gummed in a mouth, or as a secure platform to show Grandma or Grandpa your latest attempt to write your name.[3]

This stative-active conundrum about nouns exists about the word *information*. It has come to be used by most people in general, non-specialist conversations as a static noun ("Did you receive the information I sent you?"), when in fact it is a verb that needs to be used very carefully ("Were you informed by the leaflet I sent you?"). In specialist scientific settings 'information' has a very precise mathematical meaning, one that is different from the general usage of the word 'information'. It was fixed with that meaning before the New Science developed. Measuring 'information' scientifically has proved very powerful to improve the quality of remote communications, but it falls short of the meaning I shall be exploring here in ways that I shall be explaining in more depth later.

The triquetra provides a New Science informational model of the world's energy dynamics that is consistent with the systemic adaptability needed for life qualities to be renewed. Scientists who still think in terms of linear processes from transmitter to receiver rather than mutual patterning, however, may misinterpret what is said here. Likewise, in general parlance great care will be needed to understand a synergy process, because the static view of information as a noun tends to be associated with valuing the idea of possessions, such as "My shelves are filled with books", or "I have a leaflet", or "I have bought extra hard-disc storage for my information".

In the neurophysiological realm of physical nature, information is definitely a verb, not a noun. 'Formation' as a verb references the process of connecting or growing responsively into an integrated and *dynamic pattern of communications*. This pattern constitutes a form that is *worth identifying*. It has *value* enough to be recognised, even named. The verb converts into a noun, however, when our attention focuses on the potential for *impact* of that patterning form on its wider context. The formation (noun) impacts an ecology. 'It' brings about further changes of relatedness there, while retaining much of its own pattern of formation. As a consequence, a deterministic wave of impact may progress, like a cannonball through a sailing frigate during a sea battle. For example, the Red Arrows aerobatics team make a formation getting ready to inspire awe in an audience at their skill and daring; or a susurration of starlings in autumn makes a formation while preparing for migration. Memories of these experiences may influence our planning of activities on future weekends. Each formation (noun) is, however, a dynamically self-controlled process involving rich networks of communication in feedback loops between the participating elements (pilots or birds).

Let us break the word down to see its foundational elements.

The 'in-' of information refers to the quality of *relatedness* within a dynamic pattern of raw elements, which coheres as that pattern when other environmental features change.

The '*-form-*' of information is the pattern of the raw elements involved in that dynamic process of organisation. The pattern truly is a noun, properly called the datum or data, except that in the triquetral cosmology behind what is being said here, that pattern is constantly moving within. The pattern, if transmitted, will transform a receiver as it responds to the new input. However, taking radio waves as an example passing through a human body, an informational pattern (the plural noun 'data') when transmitted might pass straight through a transmitting medium with no impact. At the opposite end of the transmitter-receiver spectrum of informational processes, that *form* of data may be so densely organised that it destroys the receiver, such as an electrical surge of power conveying an accidental overload, or even a bullet conveying hatred. In these extreme cases where life ceases, there is no true informational transfer into the receiver, because data overload has brought that transmission to an end.

The '-ation' of information refers to the dynamic *change* that takes shape among the raw elements internal to the form, in which movement is transferable to some other form as its transformation. If there were no internal change, there would be no potential for transfer of information (as a verb). An unchanging transmission (static) could not be received. It would not, in fact, exist.

This dynamic view of information as a verb is pivotal to the way mind and matter can be brought into a dual-aspect unity. Seeing information as an *internally dynamic pattern that is transferable within a wider environment* can overcome the various analytical dualisms that will be described in Part III. This is because any dualism requires (or creates) an unchanging essence to be its core assumption. That *unchanging* essence cannot, therefore, informationally communicate. In theory, therefore, 'it' does not exist.

I shall emphasise this pivotal point by preferring to use the phrase *informational process* rather than the ambiguous word 'information'. An informational process adds a new dynamic relational frame into any stative noun-type concept, such as a bookshelf. Relatedness is core to the informational process that first puts another book onto the shelf, which then empowers the person intermittently to access its pages and be internally transformed by its text and diagrams. Relatedness is not a feature of a static state of possession of an unread book. Relatedness is knowable as the changes of form that are the informational processes between transmitter and receiver. Different receivers of an informational process may have different relational frames of responsivity, and so react differently to the text when reading it. That's what makes relationships interesting.

Informational processes link your brain and immune system

Your brain is NOT a repository of self-existent information. Your brain and my brain are pattern-recognising organs that can mutually transform informationally. They memorise environmental *processes* beyond your body. Memory traces are data (noun) stored as if in a memory stick, which can be *informationally* reactivated into mental processes to add their relational frames to other incoming sensory data. There is usually some error bias in this process.

It may surprise you to hear that this process of informational reactivation is equally true of your body's immune system.

Your immune system is also a pattern-recognising organ, detecting your own physical molecular biology. During the first 13 years of your life the thymus gland learns to recognise *your* macromolecules. It memorises the physical *you*. It then can identify any biology that is 'non-you', which is a potential threat to your healthy self-integrating life, and eradicate it.

There are more cells in the immune system distributed through your whole body than there are brain cells. These billions of immune cells are talking to each other biochemically about how well you are doing in the world, and whether perhaps you need to take some time out to heal, or whether now is a good time to take a few risks and 'get a life'. The biochemistry of *self-organisation* at this level is the source of your personal energy levels. It is also a rich informational process that contributes to patterning your personal identity. Biochemically, your brain and immune systems communicate with each other. Your immune system learns to recognise patterns of self and other at a molecular biochemistry level, in a similar way to your brain learning to recognise helpful and unhelpful patterns of sensory activity in the wider physical and social ecologies. These two parallel processes contribute to generating a sense of self and other in your lived experience—your unique personal identity.

It is a curious fact of embryology that your nervous system, immune system, and skin all differentiate from the early embryo's surface layer, called the ectoderm. The ectoderm is the outer of three embryonic layers—ectoderm, mesoderm, and endoderm. Together these make a hollow embryonic tube.[4] The outer layer interfaces with the world and its social ecology, which for an embryo is everything. *Pattern-recognition* is therefore a feature of interface, both physically and socially. The inner ecology of a developing human being and the outer ecologies of the environment are informationally one throughout a human being's life. While bathed in the womb, this interface is an undifferentiated relational fact of growing a life. The foetus is developing increasing potentials for diversification and differentiation. Later, when grown sufficiently to walk and explore after birth and on into adult life, the informational unity of the internal and external ecologies tends to be forgotten. It can corrupt into a dualistic informational divide, of an adult mind from its troublesome body.

Informational unity above and below the orientation horizon

This artificial divide may emerge as above and below the orientation horizon, which therefore can be rationally and safely bridged. Above the horizon, an orientated view of bodily movement in the world and in conversations includes noun-like concepts constructed from changing patterns of relatedness. Below the horizon, pre-orientation awareness as an individual does not localise experiential events in space and time. Changing patterns of relatedness can be appreciated informationally for what they are. Above the orientation horizon, as shown in this repeat of Figure 3, the world is shaped by mental constructs, which brings safety, stability, and rational social responsiveness.

Below the orientation horizon at this inner heart level of human experience, qualities of relatedness in stative or intra-active movement can be directly experienced. This makes the primary inputs to the principles of organisation more readily knowable, but without access to rational interpretation. They may therefore seem overwhelming, or quite literally disorientating, or perhaps spiritual.

In everyday practical experience, *changing patterns of relatedness* simply means, for example, that when an arm unexpectedly knocks against the edge of a table that had not previously been noticed, the presence of both the table and the arm become parts of a person's experienced reality. Having had that experience as an event, I now have a special inner ecology memory of that arm being 'my arm'. It has a different relational frame from 'that table', which lacks any previous resonating memories.[5] The presence of this experiential interruption to movement is memorised into a relational frame pattern by orientating it *with other senses*, giving 'it' an ecological context. Myriads of these patterns accumulate over a lifetime, adding unique qualities to each person's ongoing experience of life in the present moment.

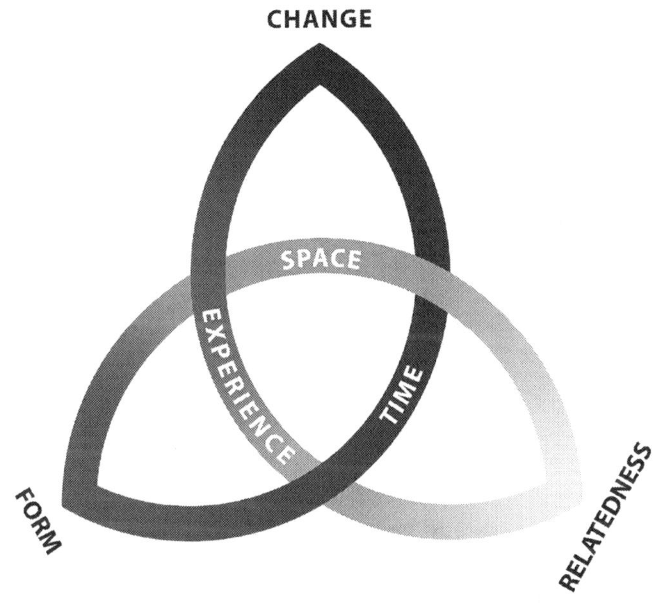

Figure 3: Three primary constructs for orientation

The New Science uses highly advanced fast computers to map the adaptable repatterning of any type of relational system mathematically. The principles of organisation are common to all systems computationally. The triquetra in its networks has the informational advantage, however, that diagrams and 3D-printed models of them make that mathematics *visually accessible* to anybody. Inner organisation of movements becomes externally visible in ways that can be remembered and understood, and used therefore to improve exploration and learning from life. The double triquetra of Figure 5 (see below) is the core icon of the organisation of movement within any object or concept. As we shall see, this can help to bridge the artificial divide between mind and matter with an informational strength. Heart-level emotions and values can also then join in partnership with reasoning for more constructive and orientated capacity to make wisely intuitive decisions.

Let's explore another example in depth. Imagine a quantum physicist in a lab, who is interpreting photographic plates from the Hadron Collider. She or he may see curving traces as a new experience. This focus, in its work-lab context, will add to the existing relational frame of that physicist, which 'says' in an orientated way that this must be a particle rather than a wave-field. However, in the pre-orientated reality below his or her orientation horizon it is an informational flow feeding back from the physical and social ecologies of the laboratory. That lab is an ecology. It was designed and built as a purposeful environment by the conversational cooperation between many people who all think in the same ways. Their agreed purpose needed a very large high-energy measuring instrument to loop out from the lab for 27 kilometres under the mountains through two neighbouring countries to play with the presupposed existence of particles and waves. That is the relational frame within which a quantum physicist will interpret this informational feedback and orientate it into his or her inner ecology. However, below the physicist's orientation horizon, where all humanity is in feedback unity with the physical and social ecologies, the triquetral model might allow a more informational interpretation of the changing patterns of relatedness that have generated the image on the photographic plate. We shall go on to see in Part IV how it can be interpreted in terms of informational synergy that is creating wave-particle indeterminism.

Personal wholeness involves constant feedback movement and anticipation. Relational patterns constantly change at an inner heart level. In so doing they inform our living intra-action of conversational engagement in life. We can know we are *part of the known*. This balance of self and others, as intra-active personal identity in the world, emerges as two informational feedback processes combine in the core brain, both having been shaped by the same principles of organisation. It is an informational process that some, if not many, people believe could continue even without the structuring provided by the brain's living tissues.

Personal identity is not 'located' as data anywhere in the cerebral cortex. Personal identity is reawakened each morning via *anticipation patterns* that have become embedded in the core brain. Social feedback loops through the wider physical and social ecologies of life collide with our personal anticipation patterns and hopes. This quality of relatedness sparks emotive and attention-seeking responses from unexpected informational mismatches in the inner ecology.

Who is 'a person'?

I have mentioned that it is a *person* who becomes orientated. It is me the person who seems real, or may seem unreal when things are informationally out of balance. What do I mean when I use that word 'person'? The relational frame of a person will become more and more important as the theme of this book develops.

The term 'person' came from the Greek *prosopon*, which was a mask worn by actors to indicate the role they played (the person) in a drama alongside others.[6] It is thus an innately *relational concept*. Being a prosopon includes bringing the audience into the story along with the other players, identifying features of the self with the different roles. Together they tell a moving story of how different people, different persons, tend to intra-act in demanding situations. The story reveals a person's inner tendency to make certain choices, revealing their character and the consequences of those choices.

Another Greek term that has been used to describe a person's relational nature is *hypostasis*, meaning 'substate'. A *person* is a *dynamic substate of a greater whole*. The substate is intra-active, having a choice-making role while continuously interacting relationally with others in their roles. Terms such as 'father', 'mother', 'sister', 'cousin', 'friend', 'employee', 'citizen' are all innately relational and could be considered to be roles within a greater social ecology. For any of these relational frame descriptors to be true there has to be an 'other'. You cannot be a citizen, for example, without there being a state to belong in. You cannot be a friend without some other person or feature of life to be responsive with. 'Person' is a similar descriptor, but generalised to indicate a unique individual *of a people*. The term 'person' implies a mutually responsive social ecology, not a private, inner, inviolable state of 'right' to be an individual. The theme will be developed that the term 'person' includes a choice-making role that influences the wider physical ecology of life (the land) out of which the social ecology emerges.

A surprising consequence follows. A whole person is incomplete alone. Each human person has their unique individual personality. He or she (in whatever way gender is defined) may feel complete alone or believe that is true, because an individual woman, man, or child may clearly make choices in isolation from others. But that individualistic approach to life is not the same as developing personal wholeness. *Responsiveness* with others and nature is vital to being a personal substate of something greater. Acting upon a potential for responsiveness is 'life-giving' for the developing personal identity. Our lives are physically embedded in networks of personal relationships such that listening, hearing responsively, and making small compassionate acts are the currency of life. Personhood means give and take within a wider context.

Conversational orientation needs feedback

In the core brain, the meeting between the social feedback loop and the anticipation pattern established by cortico-limbic feedback loops sets the interactive scene for an innately conversational view of personhood, as *informational responsiveness*. Above the orientation horizon the self-organising person gains a relational frame of being an active, effective agent making choices materially in space and time. Below the orientation horizon, that same person can be responsively interactive with an intuitive imagination, making their own interpretations of experiential events, whether as mental or spiritual phenomena. The triquetral principles of organisation of informational flow apply equally to both interpretations, and to both sides of the orientation horizon.

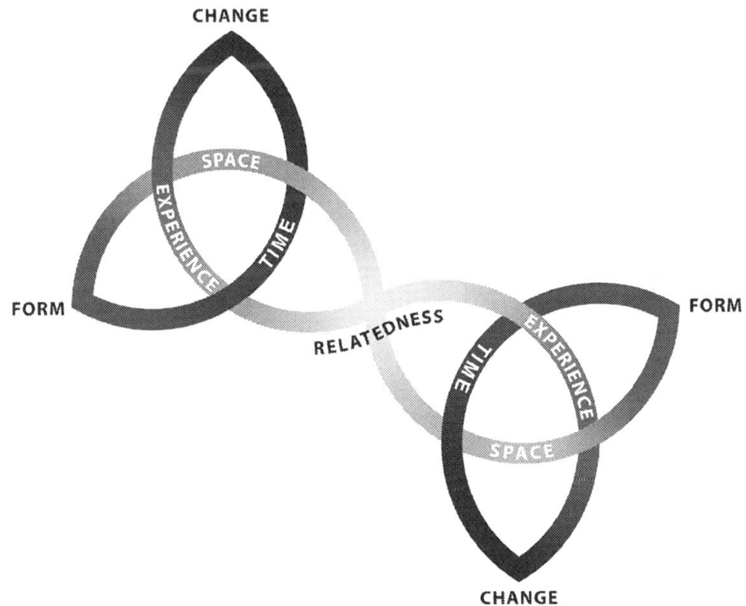

Figure 5: Conversational orientation in life

A *double triquetra* iconically shows the informational flow of a conversation. Figure 5 is therefore the core icon of triquetral cosmology, not the single triquetra.[7] The cosmology being presented here is relationally systemic. A conversational orientation is therefore required to connect and engage ecologically.

Each triquetra here represents an orientated person as an individual, but it could equally represent the meeting in the core limbic brain of inner ecology anticipation values and social ecology feedback. The relatedness principle of movement in life, in whichever context, is fulfilled in the double triquetra. An isolated triquetra directs that relatedness principle inwards only. Its reflective inward turn will remain a potential ligand of communication to turn outwards and connect an informational flow responsively with others, which will be experienced as a transformation of the inner life of an individual into a responsive whole person. In the double triquetra, iconically, each person is shown to have an 'other' who contributes to mutual existence.

Let us assume that the two triquetrae represent mother and baby, for example. The baby on the lower right experiences a discomforting change, which it (he/she) communicates through the yellow emotional attachment ligand by crying. On hearing this, the mother changes internally and transforms her previously planned time of rest to get up again and prepare to feed by drawing physically close to the baby. That changed quality of relatedness is experienced by the baby, which changes the cry into a transforming suckle search. *Somewhere out there,* the baby thinks wordlessly, *there is food somewhere nearby.*

Growth happens slowly, and the nature of this changing relationship transforms throughout the various ages of life depending on the relational transactions and types of communication that are appropriate to that age. Words come into play at some point, and technical languages then develop as people diversify and specialise in their relational groupings.

However, the core transactional processes, by which ectoderm detects patterns in wider ecologies and organises responses, mesoderm restructures and reshapes preparatory movements for behaviour, and endoderm supplies the energy in usable forms and for storage, remains the same throughout life. Although anatomically restructured over the decades, the informational flows remain iconically the same.

Informational features of the double triquetra

By carefully looking at Figure 5, we can draw out some vital core features of the principles of organisation of life's movements. In Part IV these will be shown to have some remarkable consequences for an informational way to understand how a cosmos can have personal life emerging from the air, dust, and water of the land. It may be easier to identify these features in the computer-assisted design (CAD) versions of the double triquetra, which were used to make the 3D-printed model of triquetral cosmology to be discussed later.

Figure 6: CAD version of the double triquetra

Figure 6 shows the same conversational orientation. Carefully tracing the lines within this icon will reveal a remarkable feature. Starting at the top, the blue-purple-red loop (dark to mid-tone in greyscale) passes *behind* the red-orange-yellow loop (mid-tone to pale in greyscale). Now compare it with starting at the bottom blue (dark) corner. Follow that blue-purple-red loop, and you will see that in the second triquetra it passes *in front of* the red-orange-yellow loop. A vital feature of informational flow for conversations can now be seen in the triquetra's dynamic geometry. Figures 7 and 8 show this more clearly for comparison.

Figure 7: An l-spin triquetra

Figure 8: A d-spin triquetra

The blue (dark) corner at the top represents change. If you imagine the triquetra had once been the circumference of a circle with two other equidistant points on it (yellow representing the principle of relatedness, and red representing the principle of form), then a triquetra is basically a 180-degree spin of these two points around the *change* corner. But the spin can go in two opposite directions to make two equally valid versions of this eternal knot. This intrinsic movement can be either clockwise, with the purple loop passing behind the orange one (a left spin), or anticlockwise with the purple loop passing in front of the orange one (a right spin). Both directions equally translocate the two other features of movement a half-spin. To keep the terminology consistent with molecular chemistry (where all molecules have internal spin), left spin is *levo-*, or l-spin (clockwise); right spin is *dextro-*, or d-spin (anticlockwise).

Having identified this core diversity united as dynamic features of informational movement, these two types of informational spin can be seen in Figures 5 and 6. Looking more closely at Figure 6 now, the top left triquetra representing one unique person is an l-spin triquetra, and the lower right triquetra representing another person is a d-spin (or this could again represent inner ecology meeting social feedback). Because of this diversity across the double triquetra, the blue-yellow-blue informational ligand weaves a sine wave as an exact counterpart to the red-yellow-red weaving its sine wave back in the opposite direction. This is mutuality. It represents a vibrational resonance of responsive two-way communication where diverse people, or any two self-organising dynamic systems, meet conversationally. This is synergy.

A synergic model of the reality behind the obvious

As one person communicates verbally and subtly or silently with the other from their unique state of individual orientation, the other receives and internally transforms in response to the action or idea. This leads to an inner experiential reorientation in a short period of time. In turn this inner transformation transmits conversationally back. I am not ashamed to call this mutuality the dynamics of love in its widest possible sense of enduring, life-enhancing responsiveness.

This synergy model raises the intriguing possibility that people may become informationally attuned to each other, harmonising their lives below their orientation horizons in words, deeds, and wave-field-like synchronicities as they develop a mutual understanding of life's experiences. Above the orientation horizon during this informational exchange, each will uniquely be partially integrating an orientated awareness of their wider physical and social ecologies into their mutual informational exchange. However, during a truly interesting conversation, people can lose contact with that mental environmental frame in their sensorium, and enter an informational realm which I call 'below the orientation horizon', although others may prefer to use the descriptive language of rising up into a higher realm. This feature of diversity is foundational to the personal truth of being substates of a greater whole in which one benefits the other. One person becomes an ecology for the other in a synergic conversation.

***

Returning for a moment to noticing relevant details in the iconic isolated triquetra, the semicircular loops make an endless internal dynamic that also curves in a sine-wave 'flow' of internal movement between the three corners. In both l-spin and d-spin triquetrae, the alignments behind or in front of a crossing loop are the same *at all three corners*. Therefore, the spin of any triquetra can be identified at *any* corner by carefully noting whether the loop descending to the right from any point passes behind the first criss-crossing loop (l-spin triquetra) or in front of the first criss-crossing loop (d-spin triquetra). If you can become familiar now with this feature of triquetral spin you will find it helps later to gain profound insights into the structure of matter described in Part IV. This affects the diverse and substantial ways that life may be renewed where all three ecologies meet.

Harmonisation and variability of feedback

In a shared conversational space-time-presence, the informational flow may take unexpected turns. This is because, after the internal reflection at either end of the iconic double triquetra, the feedback to or from the other person is somewhat unpredictable. This is a highly significant feature of relational personhood. It needs a pause for thought to be appreciated, so that it can be generalised later.

The internal reflection in the quality of relationship shown in Figure 5 includes generating an internally orientating mental context of time, allowing timing. Time chosen for reflection generates *indeterminacy* of the outcomes in the conversation, because the reflection corners in Figure 5 (*change* and *form*) input variable levels of responsiveness. These corners here may represent inputs to the sensory association areas from each person's cerebral cortex, where many other inputs from elsewhere in the neural networks of the brain also enter the association areas. These wider networking inputs will prime the responsiveness at each inputting 'corner', affecting the conversational orientation. As a consequence, in the conversational exchange between the two people (or systems), what is heard, understood, or received from the transmission can vary from person to person.

Feedback loops in neural networks thus create widespread *personal states of indeterminacy*, but also of potential for harmonisation. Options for choices and decisions are influenced by prevailing patterned states of activity in these cortical networks, some synergising while others are disruptive. This dynamic patterning *is* the *character state* of the whole individual. Importantly, because a whole person is incomplete alone, the character of their choices contributes to wider states in the social ecology. All the socially networking conversations have their own sets of indeterminate effects because of unique histories being memorised in the neural networks that filter informational exchange. Harmonisation at any and every level of our place in our cosmic social home, and conflict or disruption, feeds back into the core limbic brain to collide in its filters with a person's anticipations and hopes.

In summary, widespread states of potential harmonisation within systems also introduce variability into the informational feedback of communications. They allow diversification to be a dynamic possibility via an unpredictable balance between inputs at the three informational points in triquetral feedback loops. They allow *stable patterns* to form (attractor states) and *progressive flow patterns* (wave states). Both are types of patterning occurring transiently within the informational unity among intra-active substates of a whole system. This system could be a cosmos in which conversational people mix their inner ecologies as they share in the land.

The emergence of internal awareness

Most significantly of all, let us consider again only an isolated, self-reflective triquetra. If a steady state of stable harmonic feedback has become established between *any* two zones (iconically 'corners') in the anatomical structural arrangement of informational synergy (for example, the bootlacing between frontal lobe motor planning and a sensory association area), the 'third corner' of this informational triquetral system could introduce variability to that resonance (transmit), or receive informationally from that resonance. If now that receive-transmit synergy influences wider networks, the informational flow from that initial 'observed resonance' could amount to an internal sense of awareness of the presence of inner changing relatedness. Such awareness could impact personal choice to influence, or to 'stay with', the observed resonance. In the example given above, a resonance between the frontal motor-planning area with a sensory association area could be modulated in this way *by activity within the deeper limbic brain via cortico-limbic feedback loops*, and vice versa. This may represent a foundational structure for personal free choice based on personal values to arise in the inner ecology, which we shall explore more in Parts II and III.

With the interplay of three zones of the brain, each has its own type of internal function, which remains healthy while it remains in continuous responsive communication with the other two. If this threefold system becomes unbalanced, however, problems will follow. This all contributes to the power of choice that an individual person has over internal informational flow and that among a people. A person may realise that their internal choices do not somehow improve their inner coherence, and may come to feel unsafe within themselves, generating anxiety states that further complicate the entire inner and social systems as people share in the land. Healing of this unsteadiness can occur by gaining insight into the foundational stability of triquetral movement and conversational relatedness everywhere. Recognising also that there is an orientation horizon to mind, with also a heart-level depth of relatedness, can stabilise a person within the living unity of shared movement. That capacity for insight develops the power of choice for growth into a balancing maturity of character.

Time and the triquetral 4D phase space

The indeterminacy of life's events in the social ecology therefore have their roots, other than only the inner ecology. Life is not all in the individual's mind. Life is in the socialising balance of inner hearts sharing in the land and bringing free choice into mind. A Mexican Wave in a football crowd,[8] for example, emerges as individuals choose to demonstrate their responsiveness as substates of a wider people. It becomes a wave, and not an event when everyone simultaneously stands, because *time* is added to the spreading 'conversational intra-action' among people's experience of being in a crowd. Time, and timing, converts the open 3D space of the football stadium into a 4D *phase space* for variable movement.[9] A phase space maps the range of possible states in a timeframe that a dynamic system can get into. A practical example of this could be to consider the inner mental ecology of a single football fan, who takes away memories of the game that may be different from other people's, and then mentally plays with these memories as they consider alternative possibilities of outcome. This inner reflection of options, variants of the memory, takes place in the phase space of their informational minds. Memorised data from the live spontaneity that they have experientially witnessed inputs repeatedly back into a motor-planning system, which imaginatively anticipates different options to choose from. The choice of option most enjoyed depends on the individual's core brain values getting reflected back up as another modulating variable, as if from a third corner of a triquetra, via cortico-limbic loops, creating a never-ending inner game.

Let's extend this line of thought a bit further. This self-isolated individual could be represented by the triquetral icon shown in Figure 3 as an open *4D phase space of possibilities* for changing patterns of relatedness moving among their memories. The double triquetra in Figure 5, however, is an icon of a conversation, perhaps between two local football fans who have cheered their team on at the recent match. Each of them is a 4D phase space in their inner ecology of body and mind. Around a friendly drink they swap stories of the game and the referee's decisions. In this conversation, an extra level of informational movement is introduced to their 4D phase spaces.

Later they are joined by some more friends who were not at the match because each had something different that they needed to do. Figure 9 shows how the double triquetra can extend into a group identity where each unique individual fills out their informational identity as a person and as a group that is open to welcoming new friends to join them. Stories about the match pass around the table, some conflicting or disagreeing about events or their significance, which raised the curiosity and humour of the group. Eventually they go their different ways, but in their hearts remember the good feel of each other's presence that evening.

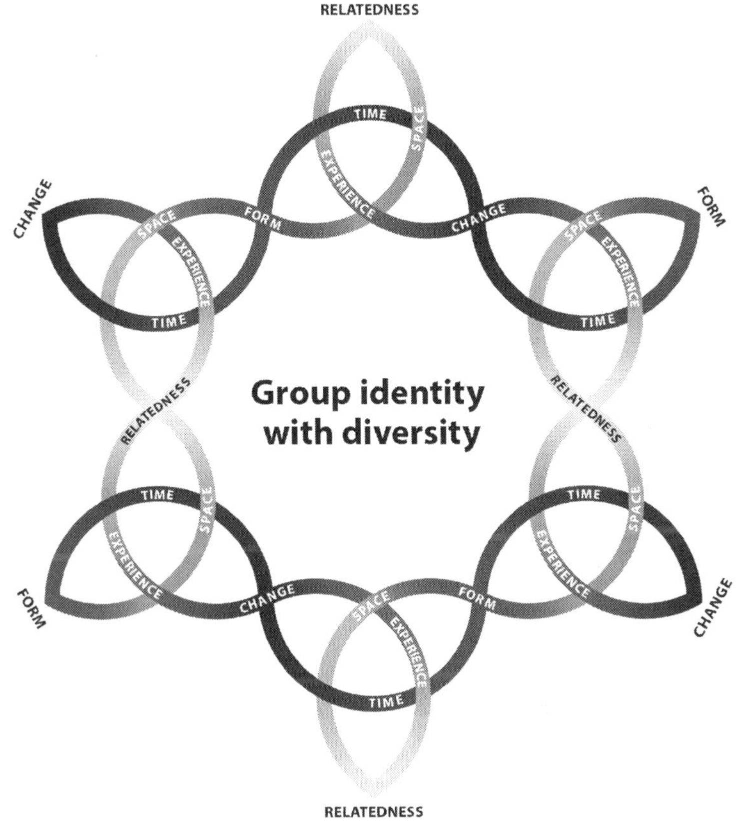

Figure 9: Group identity allowing diversity

The synergy shown in Figure 9 can create a *stable domain* of group identity. Those who belong in the group may settle into shared habits. In this way, anticipation patterns are easily and quickly fulfilled, so that less disturbing emotional energy is generated by unexpected events. Stable domains become relatively low-energy attractor states, where stative relational synergy keeps life harmonically simple and enjoyable. However, mostly people also need to get out into the unexpected and 'get a life', seeking stimulation by exploring new experiences and relationships. Finding the balance between progressively seeking change, and maintaining stable domains to return to rest, is a constantly variable and indeterminate feature of life for each person as they approach their different environments each day.

Fractal variability within a phase space of cyclical changes

For physical scientists, there is an *indeterminacy* feature in physical nature equivalent to that just described among the different brain systems of the inner ecology and in the associated social ecology. In the telescopic astronomy scale of orderly movement, this indeterminacy is known as 'the three-body problem'. Unpredictable behaviour can be seen when three large (massive) bodies (stars, planets, or moons) are able to move in relation to each other under the influence of their own gravity. The emergent phase range of movements of these huge masses is unpatterned, complex, and unpredictable.

At the opposite end of the physical ecology scale, an equivalent indeterminacy can be found in the quantum mechanics of, for example, a helium atom. Its nucleus and two electrons interact according to an 'inverse square law', which is the same law that applies in the astronomic scale to gravitational attractions.[10] No precise mathematical solution can be found to account for these complex but averaging movements at both ends of the physical scale.

Intrinsic indeterminacy therefore pervades all three ecologies, physical, social, and inner. In Part IV we shall look at how predictable, deterministic behaviour does emerge in physical ecologies, and less so in social and inner ecologies. It is in this reality that a person is an intra-active substate moving all three ecologies simultaneously by the character of their choices. The capacity for choice is held *in the relatedness between* the knowing brain and the known world. Qualities of relatedness are most actively held in the dynamic states of the core limbic brain. The quality known as *belonging* is held there, at a heart level of awareness, rather than in the orientated known world, where cortical analysis is more likely to disrupt it with options for choice than enhance it. It is in this depth of heart-level reality that an informational view of personhood can be fully developed. To unveil that depth, when it has been hidden for so long by a dominant neocortex, three concepts will make a new and vital foundation to stand upon:

- Information is a relational verb.
- Movement is the changing relatedness of patterned forms.
- Life is informational movement.

In the next chapter we shall look briefly at the informational content of a person's spoken language. To conclude this chapter on an informational view of personhood, some neuroanatomical details are provided on how reverberating harmonies can be set in motion in the human inner ecology. These are the precondition for insight to improve the power of choice.

A brief neuroanatomy summary

It is worth summarising four steps in the sensory pathway through the neocortex that together produce the inputs to the orientating triquetra represented at its *change*, *relatedness*, and *form* corners.

Step 1. Raw sensory inputs from the various sensory organs first go via the core limbic brain's synaptic filters to various 'primary sensory reception areas' scattered in the rear two-thirds of the cerebral cortex. Here they are broken down into elemental features that are detectable by single nerve cells.

Step 2. These elemental features are projected forward into nearby secondary sensory reception areas, where various elemental features are gathered into phenomena, some of which include the change, relatedness, and form features that are needed for orientation. For example, in the visual secondary cortex, *form* is informationally constructed from elemental straight lines, angles, curves, and so on; *change* from light intensity switching off or on, or from the movements of edges and ends of lines; *relatedness* from the coincidence, for example, of two or more movements that synchronously recruit enough to activate a third neuron or a small neural network. Other sense modalities have their equivalent phenomena.

Step 3. The phenomena identified in the secondary sensory receiving areas of vision, hearing, touch, and body (joint position, muscle tension change, inner ear balance) are projected forward into the three *sensory association areas*. Here they integrate into multisensory phenomena. They may be sufficiently recognisable at this stage to activate memories and their additional associations. By differently pairing the three multisensory phenomena required for orientation, each of the three association areas contributes a different orientating construct for conversational interaction—space, time, and the experience of object recognition. This Step is shown iconically in the Figure 3 triquetra.

Step 4. Multisensory, partially orientated phenomena project forward again to the motor-planning areas of the frontal lobes where a final integration occurs to integrate personal responses. These cortically planned movements may modify more spontaneous values-based preparations for response that already are initiated from the core limbic brain's filters. Cortically initiated movements are thus framed within a sensory environment in which past memories have added unique meaning for the individual to the present ongoing social and physical ecologies.

Imagine that the surface 3–4mm cortex of the sensory association areas (the grey matter) is made up of millions of columns, all perpendicular to its surface, like an orchestra of tubular bells jostling together. Each is a micro-network of small nerve cells making a hollow tube. Each tube has six identifiable sections along its length, so that six horizontal layers can be seen in the grey matter running parallel to the folding surfaces of the brain. In the hypothesis being presented here for testing, the different pairings of extracted sensory features for orientation could *mingle* in Layer III of these cortical columns in the three sensory association areas.

There *in the column wall* they would thus create a variable electrochemical filter, like a very sophisticated radio valve or transistor. This variable filter could modulate any other sensory information that is passing *up through the centre of the columns* on its way to the motor-planning areas of the frontal lobes. Cortical columns, I am proposing, function like old-fashioned radio valves that glow in the dark as they modify sensory and memory throughputs.

Of course, I am being selective in describing this. The networks and feedback loops in the brain are fantastically complex. The point of giving this simplified picture of the process, however, is to explain this important point: *any* sensory or memory phenomenon that adds richness to life (for example, colour mixtures, musical themes, interesting textures) will enter the base of these cortical columns in the sensory association areas on their way to the motor-planning cortex, and they will leave the top of that column having been *modulated by orientating features*—in space, or in time, or as an imagined objective thing that summarises relational change into the experience of substantial presence. Whatever emerges from the top of the columns goes forward for further integration in the motor-planning areas of the forebrain. There, people imaginatively work out *how to manipulate this thing called life.*

Therefore, there are two optional routes from secondary sensory areas to the association areas. The majority of interesting phenomena go to the bases of the columns and up through the middles. A secondary route for selected orientating phenomena (change, relatedness, and form phenomena) goes directly to Layer III of the micro-network wall of the columns. More detail about the selection of which route is taken during attention-focusing will be provided in Part III.

This triquetral pairing system in Layer III of the cortical columns would bring orientated order out of an otherwise chaotic background of electrochemical activity in the nerve networks of the brain. The system can be overwhelmed in various ways, however, disrupting people's preparations for organised responses. Nevertheless, it is also remarkably robust because it is basically so simple in the way it generates adaptive complexity. Inner order can be restored rapidly when the physical and social ecologies are less stressed. During times of rest and recuperative strolling, 'coming back to your senses' is a good starting place to allow that attention-focusing mechanism to unwind, and to find a 'gap in time' in which to reflect and reintegrate.

Chapter 3

Becoming Informationally Orientated

Life is in continuous creation and renewal

Given that this orientating framework for human awareness is continuously active when awake and not daydreaming, I need now to justify my claim that the same principles of organisation of space, time, and the experiential presence of energy-matter are ongoing in the aware person's physical environment. If they were not, then consistent personal communication with unpredictable others via social feedback loops through that physical environment (which includes the living embodied land) would not be possible.

Your body and mine are dust. We each are piles of dust continuously being gathered from the fields of the earth via our foods and ointments into self-organising, living, moving forms who communicate. That dust is mixing with a large dollop of water and oxygen that is regularly entering and leaving the body, so we are really breathing pillars of moving, smiling mud.

This dust self-organises under the chemical genetic influence of DNA and RNA, breaking our food down and remaking it into a recognisably living form that has a *potential range of responsiveness*. That is an absolutely parallel process to the informational breakdown of sensory input in the electrochemically structured brain, and its remaking into a sense of personal orientation with a potential range of responsiveness.

Our genes do not just code for shape, as if for a noun called a body. Our genes are another type of informational chemical memory about the environment of *life*. They are a depository of *types of chemical responsiveness* (our personality) that have survived from past generations and become embedded in our inherited DNA. The recognisable living form that emerges from the dust alongside others moves and interacts and sleeps and wakes each morning as the same person, albeit not so bright or beautiful usually on first waking. We do not remember the millennia of history that have shaped our genetic databanks. We simply live informationally by accessing different parts of it for healthy responsiveness at different phases of our lives.

The amazing fact of human life is that two recognisably different piles of dust can communicate their meaning sensibly to each other, and know that they misunderstand each other, and can enquire with curiosity of each other to know more. In this sense we are living miracles. Our lives reveal mysteries about history and the universe that we do not even understand ourselves. Nevertheless, the fact of conversation among this living diversity empowers exploration, so that the unknown and the unpredictable can be understood from different angles, or in different perspectives, enriching each individual's personal identity.

And more... this living miracle of a socially interactive body is in continuous turnover. It is like a cloud hovering over a mountain peak, which gathers in the wind at one end and disperses at the other end beyond the peak while remaining really present in its place above the peak for its time for all to see. Every cell of your body and mine is in continuous turnover. It has a limited life, but the life of the person extends and endures beyond that local cellular physiology of dust. Even those nerve cells that manage to endure for our entire lives have their macromolecules and cellular components progressively replaced. A process of renewal is ongoing at all times, day and night.

This is the meaning I have in mind when I use the term 'continuous creation'. It is almost identical to 'renewal', meaning gathering and dispersion, receiving and letting go, all within a wider context of life in its relationally changing forms. Continuous creation is the life process behind the physical healing of injuries and of damage caused by diseases and by the immune reactions against some of them. Continuous creation is also active in the brain during the psychological healing after trauma, as people gain courage to take the risks needed to explore cautiously how to trust again in the social ecology when trust has been relationally broken. It is the same concept as that described in more complicated language as *process philosophy*, which is being explained iconically throughout this book. Using the term *continuous creation* is more pragmatic, however, and a clearer opposite to 'mental construction'. Creation involves progressive growth and organisation within ecologies, rather than fitting self-existent building blocks together to make a construction.

The picture of physiological turnover and emergent mindfulness drawn above is in stark contrast to that Cartesian dualist view (itself derived from ancient Greek objective thinking) that the body is merely a disposable mechanical carriage for an eternal soul or mind. This gives the soul or mind a binary status as an objective noun, like the body but not the body. The updated version of Cartesian dualism has moved on, from a horse-drawn carriage steered through life by a soul (the homunculus) who is 'the real person', to a car and driver in which food entering the body is like fuel providing energy to power the car's engine.

But the car-fuel metaphor would look like this in a continuous creation frame of mind. Fuel (food and data) in the car's tank is informationally converted into the car's body, seats, steering wheel, windows, engine, starter motor, brakes, tyres, *and into the mindful driver* who is choosing how to shift the indeterminate stative-active states of all of these substates of a moving car. That same fuel renews parts worn down by relationally transforming movement. The energy that moves the car comes from explosions contained within firm engine cylinders that spread their tensions through every molecule and wave field that constitutes the whole car and driver. The homunculus disappears into a driverless but mindful living car. Other vehicles sensitively and mindfully recognise this dynamic pattern of activity as a different car from them, but one that is fun to be with, or perhaps at times comes over as a bit erratic and unpredictable and is difficult to connect with. Memories of past near-misses mindfully recalled affect anticipatory patterns in these other cars, so they reflexively, without clearly thinking, give it a wider space to avoid crashing while playing hide and seek, and so on. That's life. This is a description of your mentalising social body. It is more than a robot, as we shall see when we look more closely at emotion in Part II. It is a living heart and mind organising a shared physical ecology into a process of gathering and dispersal.

Moving on from a dualistic view of the body

Moving on now from the car metaphor, young adults commonly forget how much learning has gone on in their early toddler years. They rediscover this when they themselves become parents. The child's body grows slowly, and with it the personality habits patterned by genetics, because the child's growing brain is genetically shaped overall. However, its rich *internal connectedness* develops also in response to the child's lived experiences. This is called its '*epigenetic* development'. The size and distribution of synaptic connections between nerve cells physically change as part of the learning process, which affects subsequent anticipations. These neural network connections can slowly grow and slowly recede depending upon their usage. This is an inevitable feature of intra-action, within which the character of choices made modify inbuilt personality traits to affect qualities of relatedness and attitudes towards a wider spirituality. Relational guidance from parents and other emotionally available carers builds character in the child, by epigenetically adding the child's *power of choice* to modify inbuilt personality habits.

Most adults have never heard, however, that most learning happens not so much by opening new pathways for curiosity through the neural networks of the brain, as by suppressing unhelpful ones. Children naturally have an inbuilt curiosity (playfulness) that converts into lessons learnt by developing *inhibitory feedback cycles* that *suppress* surrounding nerve activity beyond the focal point that is so interesting!

I need to re-emphasise this point. It becomes crucial to becoming informationally orientated as a mature person. It is foundational to the way informational processes convert the physical matter of the brain into a mindful process of reorganisation and renewal that includes seeding memory traces in neural networks. These memory traces can be resonantly reactivated by current neural patterned activity to become informational, and thus formative for evolving behaviour. With mental or emotional overload, the lack of inhibition overwhelms memory, and disorganises responses.

Attention deficit has become a major problem in IT-dominated societies. It is my opinion that young brains have been allowed to become overstimulated by screentime without discovering how love is also manifest in setting creative boundaries for a child's activity. Using drugs rather than parental guidance to inhibit 'surround' neurological activity is not a sustainable or humane solution.[1] However, to revive the honour in family life that used to be attributed to assertive parenthood may be difficult to achieve as a sustainable way to have an emotionally available adult consistently part of the social feedback loops that develop young minds and brains. Abandoning that responsibility to teachers and lawmakers is not a sustainable option, although commonly needed in a society brought up on screentime.

It is well known in medical settings that without self-controlling 'surround inhibition' of neural network activity, the muscles of the body go rigid in an extensor spasm that eventually prevents breathing. Such is the nature of strychnine poisoning. The brain would go into sensory overload, and possibly start fitting. The brain focuses attention by screening out noise in order to increase informational exchange. Self-organising feedback systems, which are the essence of personal life, need self-discipline rather than drugs to truly come alive. Learning to say "No!" is probably more important for informational clarity than always saying yes. Without an occasional firm "No-no!" the person does not develop an inner strength of character as part of a continuous creation. This is about cultivating a knowing brain, so that what is known about life can build social responsiveness with an inner adaptable strength.

Orientating perspectives in diverse settings

As people grow in adulthood, they interact in increasingly diverse life-settings. Various specialist languages develop to describe shared interests and to tell the stories of their explorations. If the principles of organisation remain true, specialist languages will have traces of the same core neurological principles of organisation if they improve social responsiveness and mutual understanding. Table 3.1 demonstrates how this might be.

The point is to gain the awareness of principles at work behind the neocortical construction of a *linear* language of words above the orientation horizon, which attempts to reference *patterned dynamics* below the orientation horizon that may subtly communicate with others more directly at heart level. Developing this type of insightful awareness of the informational principles of shared movement could help to overcome the problems that will be described in Part III.

Table 3.1: Three perspectives on life in different language-games

Specialty	***Specialist language***		
Neuroscience	Form	Change	Relatedness
Medicine	Physical	Psychological	Social
Scientific	Form	Function	Cause-effect
System analyst	Structure	Process	Output
Personal	Personality	Character	Spirituality
Pastoral	Physical body	Life	Relationships
Theological	Body	Soul	Spirit
Artist[a]	Wholeness	Harmony	Radiance

[a] James Joyce, *A Portrait of the Artist as a Young Man*. Routledge & Kegan Paul, London, 1916.

Demystifying the language of mathematics

Could the same principles of organisation be detected in the language of mathematics? Maths is a mystery language for many people. (Einstein's famous gift to humanity, $E = mc^2$, is the only mathematical formula that will appear in this book.) But I believe the same principles of organisation can also be found there. If so, then some people may feel they understand maths enough to worry less about what scientists are up to. Maths is a language of pure ideas. This language describes movement... *dynamically transforming connections into patterns*. One pile of coins transforms into a different pile. Are they equal in value? Maths can describe the transformational process accurately. The weird symbols used to record this language will always be obscure to many people, but mathematicians (and my daughter is mysteriously one of them) somehow can imagine and represent symbolically any number of *dimensions*, which bear no relation to the life most of humanity leads in four dimensions to orientate the known world as a space-time-reality. Mathematicians can calculate dynamics and outcomes in a sort of imaginary hyperspace (such as in chaos theory), which to everyone's amazement may then be visible in the way matter and energy really do process in this turbulent world. Pure mathematicians find this feature of 'earthing' their calculations quite irritating, as if reality misses the point somehow.

At the risk of irritating pure mathematicians, however, I am a great believer in earthing personal life, of which mathematical imagination is a human part. The point I want to emphasise strongly here is that I believe mathematicians' brains calculate algebraic equations according to the same triquetral principles of organisation that orientate every human being's real-time chaotic sensory inputs. Take as an example Einstein's famous $E = mc^2$. We know the reality of the different feel of energy (E) and mass (m) in the known world, but with clever mathematics and physical experiments we can now be certain that they are related (=) in some transformational way. The 'equals sign' indicates transformational relationship.

Scientists will see in this equation a description of the way the known physical universe works. I see in it a description of the way the knowing human brain works. They both process *informationally* according to the same principles of organisation.

Table 3.2: Triquetral principles of organisation applied to pure maths

Specialty	***Specialist language***		
For orientation	Form	Change	Relatedness
For mathematicians	Variables	Operators	Equality
Mathematics	E, m, c, etc.	x, +, 2, ÷, √, etc.	=

In Table 3.2, the triquetral principles may become clearer. Energy (E), mass (m) and the speed of light (c) are the *forms*—the *variable elements* that make patterns on either side of the equals sign. These forms repattern as they *change* in directions described by the 'operators' +, −, ÷, ^{2}, and so on. The *relationship* between transformed patterns (in an 'equation') is one of equality (=) before and after the transformation of the elements. There are other *relatedness* states in mathematics than equality, such as 'included within' in Boolean algebra.[2]

Pure maths is a language of movement and patterning. Its mysterious ways can inform the planning mind, and lead to logical anticipations. Applied maths connects the planning mind with matter's physics, chemistry, and engineering. Biochemistry and biophysics apply the language of maths into the inner and social ecologies of living systems. Therefore, informational mind and physical matter mutually influence each other, both transforming in patterns that are consistent with triquetral principles of organisation. They are dual aspects of the same informational movements.

Preparing emotionally for the paradigm shift

Recognising how the same principles of organisation can appear in specialist and foreign languages, and in the mindful planning of daily life, might encourage mutual cooperation across diverse specialisms and cultures. In this, I am preparing the way for Part IV, 'Belonging Together in a Land of Quantum Gravity'.

Thomas Kuhn observed that scientific revolutions do not happen without someone losing valued ideas or status when a new paradigm of thinking emerges.[3] A 'Kuhnian revolution' of science is happening now, having started in the 1970s. Enlightenment Science and its philosophy of *atomism* is settling into a 'special case' status, valuable for what it is, but needing to recognise its own limits. A more flexible view of the universe as quantum gravity space and time *shaping from within* is gaining ascendance. This turns the known universe inside out. The universe is a cosmos, an open system that is expanding at an ever-increasing rate, in which people's knowing minds and inner aware hearts are securely resident *and continuously being recreated*. There is now a more living basis for reshaping an ecological understanding of the knowing mind, and of spirituality (not to be frightened of this), and of human society, just as followed Galileo's day.

The Copernican Revolution in science started in 1593 as a mathematical exploration that was later confirmed experimentally by Galileo's telescope. This exploration continued over the next century until Newton unveiled the laws of gravity. This firmly shifted the earth from the centre of society's prevailing mental picture of cosmology. An emotional-social-spiritual upheaval followed. Understanding the psychology of our present quantum gravity scientific revolution may enable our generation to turn potential loss reactions into personal development gain, as people let go of old and restrictive ways of thinking to explore deeper heart-level cooperation. This opens new possibilities to stabilise our shared future with new knowledge.

The triquetral way to understand our intra-action more deeply in a quantum gravity cosmos[4] shifts mental attention from primarily valuing *enduring objects* (forms seen as static nouns), to primarily valuing liberation into *choice to enter freely into enduring patterns of relatedness* (verbs as dynamic processes). The Enlightenment Science of the last 300 years developed above the orientation horizon with a psychological assumption that material forms—objects and object recognition—in a space-time *closed-system container* were primary. To explain life in this frame of mind, change has to be added to objects, as if objects (nouns) are primary and need to be set in motion somehow. The New Science since the 1970s has transferred the focus of attention from the resulting linear view of relatedness as cause and effect, to the *transformational process* when *patterns of relatedness* (as perceived forms) mutually repattern as emergent new forms.

The positivist philosophy of objective reality is giving way at quantum level empirically to the process philosophy[5] of 'it's all change first and last, within which forms emerge and disperse'.[6]

Letting go psychologically of the *sequential* notion that change needs to be added to form in a linear timeframe of cause and effect is vital to understanding the triquetra. It is vital even more so to understanding how the double triquetra represents the nature of informational repatterning. These two icons display the *interdependent co-origination* of change, relatedness, and form *in movement*.[7]

Without movement, nothing exists in the New Science world, nor in the ancient world where the triquetra was artistically and philosophically conceived. Without including all three features co-equally in the principles of emergent organisation, there is no mindful movement among matter, only patterns of progressive corruption, stuckness, and conflict. Triquetral movements below the orientation horizon generate continuous creation and emergent life as perceived above the orientation horizon. In Part IV, a triquetral entanglement model will be described in which the views of two ancient Greek philosophers, Heraclitus ("Everything flows") and Democritus (originator of atomism and an inspiration for the new 'granularity' of space-time[8]), would find resonance. They both belong in a New Science triquetral cosmos. They are not contradictory views when understood below the orientation horizon. We need to unveil what happens there.

The veiling of changing relatedness by object recognition

But... there are barriers to overcome before we can arrive to a welcome 'at home' in the land of a quantum gravity universe. I shall mention two examples here. Both are products of object recognition, and are only barriers while that psychology remains dominant. Objects hide the view of changing relatedness when a team of people orientate to plan their paths through the known world, and through the emotional inscape of the knowing mind.

(a) By Shannon's numerical quantifying of information (noun)
Scientific 'Information Theory' differs in its emphasis from the *informational dynamic* of mutual knowing synergy represented iconically in the double triquetra. Care is needed to understand how. This becomes important later, because this veiling of core relatedness has become a major worldview in the age where ICT is reshaping the social ecology of the world.

In the 1940s, before computers existed, and before chaos theory[9] and complex adaptive systems thinking[10] had emerged in the 1970s to challenge his mental framework, Claude Shannon developed a very clever mathematical model to measure data transfer rates and corruption in telephone wires. He was aiming to find ways to improve the quality of speech and telegraph transmission. To know how much data had been lost, you need a number to quantify what was transmitted and what was received. The language in common usage then was to talk about measuring 'how much information (noun)' was received or lost along the way. These days we know that he means data loss (noun), not information (verb), but the object recognition notion that information can be quantified and measured has left serious lingering problems, as we shall see in Part III.

Scientific Information Theory based on Shannon's equations has mushroomed in its development since high-powered computing became widely available. Widespread uses for measuring 'information loss' (noun) have been found in digital communications, largely focused on measuring entropy, or loss of information into randomness. The key issue to bear in mind when interpreting this *scientific language* about 'information' is that the numbers refer to the possible range of states (patterns) that a closed system (or an enclosed system) can get into. They do *not* refer to a quantity, or an amount of information, as if it had an objective reality that could accumulate, like cash. The numbers refer to the number of patterns that a pile of coins could transform into. Typical language used by scientists is that these numbers indicate "how much information I can get from this system". 'Getting information' is a relational transaction, but the correct English grammar if the scientist wants to be truly triquetral (dynamic) rather than merely Enlightened (objective) would be 'getting informed'. "How clearly could I be informed by this system?"

Scientific Information Theory, properly understood as *transformational repatterning,* is therefore consistent with triquetral principles of organisation, but since 2011 a serious problem has arisen. Entropy is a measure of disorganisation *in a closed system,* where there is no transfer of energy in or out. But in 2011 the Nobel Prize for Physics went to Saul Perlmutter, Brian Schmidt, and Adam Reiss for their discovery of the accelerating expansion of the universe. The cosmos, the universe in which mindful awareness emerges among the relatedness of material processes, is not a closed system. It is an open system that is not just expanding, but expanding at an ever-increasing rate. The Second Law of Thermodynamics about order degrading to entropy only applies in closed systems. The mathematics of Information Theory only applies if people experimentally shut off parts of the cosmos from the whole. Here we are, back at parts thinking. Objective experimental environments are only parts cut off informationally from the whole, and therefore destined to die. How much the results from cut-off bits of a living cosmos can be applied to the whole cosmos is a highly debatable issue. This book will hopefully inform the debate.

Shannon's formulae are extensively and very creatively used to refer to data transfer rates from storage, and there is no harm in continuing to use this most powerful mathematical language to improve the quality of communications. But life is not all about possessing a mythical noun called information. The numbers of bits, bytes, kilobytes, terabytes and so on that people 'hold' refer to speeds of data transfer *from* storage, and rates of corruption of patterns in the channels between transmitter and receiver. These numbers refer to dynamics, *not* to 'amounts' of data. The 'information' is not an object in storage. Information is not a self-existent noun, for example like a leaflet that could be put in someone's hand, or a memory stick that could be posted to a friend. Misunderstanding information in that way is like holding on to and admiring a psychological fossil from the past Age of Object Recognition.

In a triquetral cosmology, which is consistent with an open-system universe, an 'informational process' means the *adaptability* within a system of relationships. This shifts the emphasis and meaning from Shannon's original purpose. Scientific Information Theory measures an *abstract* number of possible adaptations within a closed system. It does not refer to the qualitative aspect of potential learning, growth, and renewal that might occur within an open system, when feedback observation within its openness modulates localities in relation to each other (informationally). In an open system, energy transfers lead to informational adaptability and reorganisation. In an open system, the number of possible states is *immeasurable*. The constraints on learning, growth, and renewal in an open system need to be described other than by the imposed container that makes it a closed system. For example, we have already considered how extended feedback networks of triquetrae can provide an account for variable receiver sensitivity. All that is required is that the more extensive systems have subsystems that can communicate their states of activity. If a receiver is rendered insensitive by remote activity, so that it is not transformed relationally by new data transfer, then in an open triquetral system no information (noun) exists, or better, no informational process has occurred at that location in the whole. This is a requirement for process philosophy, that organisational states can come and go. The potential life of a closed system can die, but in an open system it can then also return.

(b) By images representing quanta in loop quantum gravity

The word 'quantum' simply means a small package. It has become associated with 'a quantum of energy', the smallest package of energy that can have a transformative effect in the material world. Measurement of the dimensions of this granular quantum show it is very small. The Planck length is about 1.6×10^{-35} metres, which is about 10^{-20} times the size of a proton. Some mathematical models of quantum gravity represent how a quantum of energy may interface with others and scale up into the known world by reference to a tetrahedron, or other 'orthogonal shapes'.

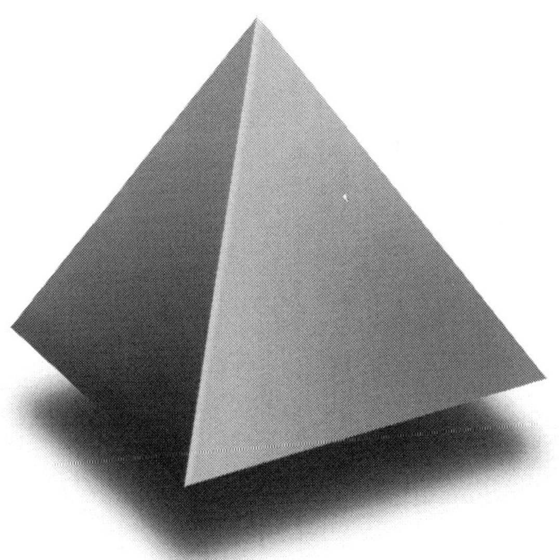

Figure 10: A tetrahedron

Unfortunately, this could encourage the unwary to make the classic psychological mistake of giving the quantum an objective spatial or structural status. A key feature of loop quantum gravity theory is that space-time is granular, like a foam of spinning energy packages at this minuscule scale, which scales up to shape the entire history of the universe in its bubbling turmoil. It is called a spinfoam model. The four sides of the tetrahedron are a best-guess attempt to represent the maths of that turmoil *as a core regularity*. However, it is deceptively alluring to think that any visual representation will activate mental associations that tend to orientate us in an object recognition frame of mind. The double triquetra may prove to be a safer, and even more expressive, way to represent those granules instead, as the change dynamics of *relational movement* in spinfoam. Movement is not added to quanta. Movement has a granular connectivity that is quantal. The triquetral model pictures that granularity as core *vibrational synergy* as shown with a sine wave in Figure 11.

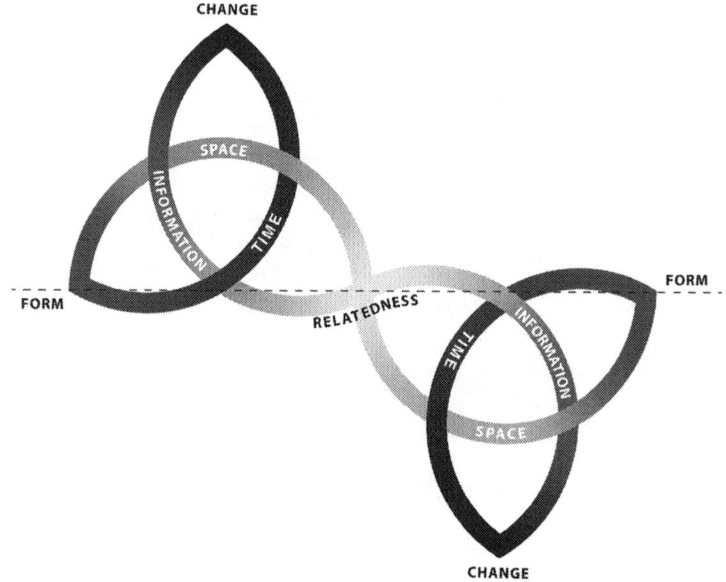

Figure 11: The double triquetra as core vibrational synergy

The double triquetra is therefore an icon of a *quantum change*. We shall see throughout the rest of this book how the synergy of a 'conversational exchange' can scale up through entanglements to create energy in wave-particles, the interactions of which make matter truly substantial. The movements scaled up become predictable enough for us as intra-active human beings to make anticipatory judgements about life, and choices for action or rest within it. The same principles of movement apply to any scale, down to the Planck length and up to the stars and black holes, and also into the depths of the human inner heart and out to the clash between civilisations and the love that can hold societies and families together. We each, when given a chance to choose, can influence life's movement on one scale or another.

The triquetra, as we have seen, has inner spin that profoundly affects how entanglements of quantum changes manifest in our human lived experience. The double triquetra is an icon of *mutual informational change*. It could be a small item of knowledge or experience transferred as an anecdote between two people, or it could represent the smallest relational change in a quantum computer, called a *qubit* of information (noun or verb?).

We shall explore in Part IV how qubits shape both the knowable world and the knowing person. In Part II, however, we shall look first at the undervalued and unexpected contribution that emotion makes to scientific exploration, scientific revolutions, and human development. Neuroscience has shown that decisions are made in the limbic brain, not the neocortex. Part II explores how the influence of extreme individualism moved the psychology of emotion away from its social ecology purpose. Emotion is primarily a *physiological* repatterning of the individual person within a moving social ecology. It has a social messaging role within groups, scaling up from the micro to the human group experiential level as seen in Figure 9, where heart-level dynamics are emerging from the physical ecology.

Emotion is not just a psychological side-effect of thinking. In a triquetral cosmology, emotion is the meeting place of our physical, social, and inner ecologies in the overall ecology of life. People's spirituality has a deeply emotive content as an anchor against the turmoil of life, which may be seen as informational transformation. Others call the same phenomenon 'thinking clearly'. We shall be looking at how much of this emotive informational communication within wider contexts proceeds below the orientation horizon, and how much above it. Relational synergies below that horizon may continuously create the personal energy that is life-enhancing. Sadly, relational disconnection can have the opposite effect, presenting in a wide range of mental and social disruptions, with harmful effects on the physical ecology affecting everyone.

But we move on then to inform the power of choice that all human beings have, so that life-enhancing decisions become more clearly reasonable when informationally reconnecting heart and mind.

Part II

Heart-Level Emotion Is Energy in Motion

Chapter 4

Mind as Consciousness, Thought, Will, and Emotion

The person as a social context for emotion

Between the macro scale of gravitational fields that warp light and time, and the micro scale of quantum mechanics, is the human mind that births all its fantastical ideas about space, time, and reality. The human mind births many fantastical ideas also about the lives of other people with whom we share our space, our time, and our reality. The world transforms physical relationships energetically; the mind transforms mental relationships informationally. Their *principles of organisation* are, I propose, identical. This allows a mind to become orientated in a physical world shared relationally with others who may think differently.

In Part I we reviewed how a person is a substate of a social system, an individual of a 'people'. A person is physically grounded in a wider ecology, and emergent in that ecology as a continuous creation with a mindful, choice-making influence. Body and brain are in continuous biochemical and biophysics turnover in that changing environment. That turnover is the physiology, the life chemistry, of an embodied electrochemical system. This living system of informational communications at every level has many inner substates, and participates as a substate in many social networks.

When personal choices, achieved by balancing these substates, alter human behaviour, the physical world transforms to some extent, either constructively or destructively. There is an informational exchange between them. If you see an untidy garden or yard it tells you something about the property occupants, something about their values. There is scaling of organisational patterns from the inner ecology of a person to the physical ecology.

The world of quantum gravity and the mind of ideas can mutually reorganise. This is only possible if they share the same informational processes. However, a mind can only emerge from the physical body and its brain if these parallel processes can also diverge from each other. In so doing, one creates a mental informational world from memories and ideas. This *parallelism of processes* allows mental exploration, the consideration of potential adaptability of response, and comparing ideas that can then be tested empirically in the physical world to obtain its feedback. It allows anticipation to bridge time, space, and reality with a parallel mental process. These dual aspects of organisation may reconnect then into a single process of a *mindful choice of embodied behaviour*.

However, the process of informational divergence and mindful reintegration into movement can go wrong. Various physical, social, and mental illnesses or problems can follow, which are not the subject of this book. This book aims to prevent them by learning how to restore the preconditions for creative life.

The mindful non-divergent state is a variable balance between stative and active features of physiology—of the biochemistry and biophysics of life. It is a mindful walk in the physical world with others. Mindfulness is currently widely mistaught in the West as simply a calming exercise, with little instruction on how to harness the clarified mental engagement into a richer social dynamic that also stabilises and nurtures the physical world. The mindful state is neither a withdrawal from the world, nor an optional activity as time out from busy lives, nor is it a static state of perfect emptiness, although all of these may be stages of learning to be mindfully fully engaged actively in our social and physical ecologies. Mindfulness is an integrated, dual-aspect way of living in movement informed by the whole of

Creation and informing the whole of Creation of your presence and personal values from your place within it.

Mindfulness meditation, if seen as a *withdrawal* process from shared life into a state of individual perfection, cannot for example compensate effectively to prevent the social impact that another person may have in society or in the physical ecology who is wilfully, selfishly, greedily, and perhaps violently having a disruptive influence. But the New Science has opened a door to connect well-tried ancient traditions of personal development with an improved knowledge of how the person can compassionately activate their real presence in the changing social and physical ecologies. The view being presented in this book sees mindfulness as a range of stative-active *states of informational participation in ecologies.*

My aim is to show how a new way to become mindful of *patterns of emotion* provides the evidence needed to name the personal values that initiate movement in active social states. Embodied mindfulness then offers a way to navigate with these values to shape life in harmony with the principles of organisation that are life-enhancing. The resulting strengthening of character can sustain informational social engagement. Part III looks at barriers to achieving this heart-level insight and empowerment of choice. Part IV will justify the claim that an embodied mindful walk in a shared life with others who differ from you establishes the foundational state of human life as the presence of loving kindness. Personal choice to renew the depth of relatedness that achieves this core active-stative state is life-enhancing.

Mindshifts about emotion

According to the *Oxford English Dictionary* the concept of mind includes consciousness, thought, will, and emotion. This definition of mind is the foundation for the informational view of the triquetral mind-matter ecology of the human being that is presented here. Much Westernised (Greco-Roman) thought for nearly three millennia has concentrated on knowledge and reasoning as the uniquely human characteristics of mind (Greek, *nous*), however. Emotion was considered disruptive of logic and a distraction from virtuous decisions, so the associated value has been that it should be regulated or excluded from an honourable, rational experience of life.

In third-century BCE Greece, for example, Zeno of Citium recommended that emotion be excluded from pure reasoning and logic to build the *virtuous civilisation*, which resonated with the earlier Platonic Idealism by which Plato proposed that a 40-year education period might be needed for leaders to regulate their emotion enough to make wise decisions when managing the Republic. Thought, will, and reflective awareness became the gold standard of the Stoic movement in philosophy that followed Zeno. This has more recently been taken as the basis for present-day cognitive psychology ('cognitive' here meaning 'types of knowledge and understanding'), which considers emotion to be merely a side effect (an epiphenomenon) of habits of thinking.

A Stoic interpretation of 'mind' is individualistic. If not seeing emotion as an experience of relatedness embedded in a social ecology, as the triquetral view does, a Stoic will view an individual's emotion as a side effect of knowledge, belief, or thought that is meaningless. It is considered a feature of human weakness, no less so in the Far East where Confucius taught, so much so that the Chinese languages have very few words to describe emotional states, which are considered to be somewhat dishonourable. To the individualistic value system, when reason is the purest virtue, emotion is intrusive. However, the New Science brings the new benefits of *adaptive systems thinking*. Movements of emergent *patterning* create order in systems (or disrupt it).

The evolution of patterns of activity requires a different set of rules, a different logic, from the codified deductive and inductive logic of ideas that philosophers have defined as linear informational chains of premises and conclusions. Patterned systems get into states. They have resistant habits. They potentially learn. They have critical points of breakdown or transition. They reflect life.

In Part III we shall see how states of conversational orientation can divide into dualistic, individualistic, and potentially linear mental analyses that limit the picture of life in the fullness of balance below the orientation horizon. Its various patterns of thinking all cast a veil over the inner heart's dynamics, altering value systems and corrupting qualities of relatedness. To reverse this trend and replace it with a New Science process picture that reintegrates these three analytical perspectives, emotion will need to be seen not in a psychological light, but as an ecological and systemic equal partner with reasoning to make life-enhancing choices. Four mindshifts about emotion will help.

(a) Emotion is physical

Currently, emotion is considered in psychology mostly in terms of behavioural dysregulation, as a linear cause of cognitive psychopathology.[1] Emotion is discussed mostly in a stimulus-response behavioural framework, as the psychological phenomenon of an individual's *feelings*. In many cultures, the words for feelings and emotions are used interchangeably. However, in the embodied mindfulness of a triquetral ecological view of the whole person, it is best to separate these two terms to appreciate their complementary qualities.

Emotions are life chemistry. Their dynamics move at the interface of all three ecologies, physical, social, and inner. They are deeper, below the orientation horizon, than feelings. Feelings are cognitive states developed in the neocortex as heart-level emotional movements connect with memories for interpretation. At the risk of saying too much too soon, in the model being presented in this book, it will be helpful to associate emotions with awareness, and feelings with consciousness. These two concepts are in similar manner complementary but different.

Emotional physiology involves hormones coordinating changes of general body chemistry through the mesoderm, and cytokines and neurotransmitters coordinating changes of immune system and neural network functioning at the ectodermal interface with wider ecologies, both of these communication systems coordinating responsive feedback cycles. Pheromones released in perspiration repattern other sentient creatures' preparation states, which allows coordination between individuals into a group in response to anyone's preparations for change. In the inner ecology that is sensitive to and responsive to the social ecology, every intercellular communication system is transformed in these emotive responses.

On seeing or hearing something shocking, the rapid release of adrenalin and cortisol into the bloodstream and via autonomic nerves will change the chemistry of probably every cell in the body within a second from normal maintenance mode to survival mode. The key concept to grasp here is that the emotional 'mode' of chemistry in all the body's cells is a coordinated pattern. It prepares the whole person to make a survival response—fight, flight, or freeze (and some people add 'fellowship' to emphasise how adrenalin helps people to talk their way through problems or to project voice from a stage to an audience). The pattern for anger is different from the pattern for shock. So also is the biochemical and biophysics repatterning of embodiment for guilty self-questioning, or euphoric denial, or depressive emptiness having seen one's limits, or the joy of meeting a loved one. These emotive modes prepare different behaviours rapidly. They are repatterning preparation states that can then be modulated or regulated by secondary processing in the neural networks of the cerebral cortex, which adds character to the emotive personality of the individual person.

The fact that the neocortex can regulate the core brain's emotive preparations follows only because those emotive preparations have first informed the neocortex that social ecology changes have challenged the personal values that are embedded there, in the core brain's sensory relay stations. The emotive preparations are pre-orientation; but they are *not* meaningless. The opposite is true. They are vitally informational, arising from personal values in a social ecology. If an attitude of regulating or excluding emotional experience descends from a cortical analysis that has been insensitive to emotive information, then the descending regulation may be *valueless* (in the triquetral understanding that values are embodied in the core limbic brain's sensory filters). Emotion and reasoning have different roles in decision-making that complement each other to make the decisions life-enhancing. They are equal partners, vital for mindfully living in social environments.

(b) Emotion is the social messaging of changing inner states

As mentioned in Chapter 1, an emotional preparation state includes facial expressions, body language that others can read, and variations of tone of voice called paraverbal communication (the volume, pauses, tone, sing-song variation, musically rhythmic feedback in a conversation). Paraverbal communication is almost entirely driven by emotion. Eighty to ninety per cent of the meaning of a spoken sentence is carried in the paraverbal intonations. There is probably an infinite range of ways of saying "No", some of which mean "Yes!" below the orientation horizon. Emotions also perspire chemicals, called pheromones, that informationally convey basic physical states to other sentient creatures so that they too can prepare for a change of action. Emotion in its unregulated state initiates shared movement, although civilised people mostly try to hide it. The primary purpose of emotion is to relationally knit together sentient creatures to survive and thrive in an unpredictably changing and competitive social ecology. Human beings are no different from other creatures in this respect. We simply have a more complex neocortex that can add at least as many extra problems as it can orientate solutions.

Emotive messages have survival value across species. We know when a dog is angry, and we know when a dog feels guilty, and we can regulate our responses accordingly. Emotion can provide the heart-level strength to face changing situations reasonably and thoughtfully. The balance achieved gives people character.

(c) Emotion is preparation for movement towards or away

At its most basic survival, thriving, and reproductive level, emotive changes of the patterning of whole-body physiology prepare people to move. They prime the personal energy patterning for movement either towards or away from social or physical situations. These movements occur in relation to personal values embodied in the core limbic brain's sensory filters.

In an experiment conducted at a university, the seats in a lecture theatre were sprayed in a randomised pattern with male and female pheromones. When the students entered the hall, there was a high correlation between male gender sitting on seats sprayed with female pheromone, and female gender sitting on seats sprayed with male pheromone. When the students were asked why they chose those seats, there was a mystified response with a typical comment being, "I don't know. It just seemed right."

This is an example of subtle emotive communication below an orientation horizon guiding orientated movement. It is happening all the time. A vast range of subtle emotive communications using biochemistry and biophysical fields is probably influencing the choices that people make moment by moment every day.

(d) Emotions move an integrated adjustment process

The Doppler Effect on the sound of a train approaching and then receding from a person standing near the rail track is analogous to the way one emotion can transition to another as situations change. The pitch of the train's whistle or horn changes with the relationship to the danger approaching or receding. So, too, the emotive patterning of whole-body physiology can change its preparation state as a situation evolves, whether a situation that challenges personal values, or one that resonates constructively with personal values. Tense anxiety, and subsequent relaxed relief, is an obvious example of changing emotional energies on receiving good news. It is brought about by informational exchange. These are not random emotional states. They are not meaningless or signs of weakness. These emotions are informational resources *within the reasoning mind and heart* about how personal values are repatterning the person's inner heart at a pre-orientation level within wider changing ecologies.

Repatterning is not a linear process, however. There may be a time sequence to behavioural preparations for adjustments, but various overlapping timeframes arising from different values being challenged along the way can affect the evolution of complex emotional preparation states. For example, socially moved healthy physiology in a friendship can be rendered chaotic suddenly by an inner psychological influence as memories of some former traumatic events become activated. Also, patterns in the physiological preparation states are not driven by one hormone at a time. A dozen or more hormones change in their balancing of effects, inputting informationally from different subsystems within the whole person. Feedback occurring at every level of the inner and social systems also has its effect, whether stabilising or destabilising.

The New Science now provides the concepts and tools needed to map these complex interactive processes. Once these patterns of emotional preparation states are mapped, dynamic as they are, people can make scientifically logical predictions of probability from them. Over several decades in the hospice movement and in other settings, research into the emotional adjustments to the loss of valued people and things has identified core sets of emotional states that together can now be seen to make a healthy *adjustment process*. In this process of movement from shock, where people doubt their ability to manage a situation, to developing new abilities to come through difficult times stronger, one emotional state evolves into another to prepare the person to move in different ways to navigate the situation. In Chapter 5 we shall consider this set of healthy adjustment emotions in greater detail. When understood as parts of a single, integrated process, emotions that might otherwise have seemed to occur randomly gain instead a meaning and a useful purpose as parts of our relational humanity.

***

Summarising what has been said so far, emotional preparations for movement are the meeting place of the three personal ecologies.

- The physical ecology is active in the electrochemistry of hormone and other communication systems at cellular level.
- The social ecology is active in the informational movements of relatedness that stir responses at a heart level.
- The inner ecology is active phenomenologically as psychology, in memories, anticipations, and the hope-and-fear-based patterning neural activity that initiates reasoning, planning, escapism, or worship as people

think and reason their way to constructing orientating options for response.

The inner ecology thus includes constructing as wide an overall mental picture of life as is possible to integrate lived experience. Most of the time, mental integration will require that some heart-level awareness has to be excluded from consciousness to maintain internal consistency in the demands of everyday living.

Thinking logically about adaptive system responses

Science is based in inductive logic. A range of specific facts or observations is gathered from controlled experiments, compared and contrasted for relevance and reliability, and then a general statement is made that is consistent with them, which is assumed may apply to other similar situations. For example, "We have observed a thousand apple trees in a hundred countries, and in every case the apples fall to the ground at some point. We therefore propose that there is a force pulling downwards towards the ground that we shall call gravity, and that this force will also operate on objects other than apples." Inductive logic moves from the specific to the general. It is sometimes called bottom-up reasoning. The general statements that follow are *probabilities*, or hypotheses that can inspire further exploration and testing. They may be wrong. Inductive logic never *proves* anything, although popular imagination may hope that science offers proofs and certainty. It does not. The inductive method can only disprove a hypothesis by making an experimental observation that is inconsistent with it. Even then, proponents of the hypothesis will critique the experimental control situation, and refine the hypothesis to account for the misfitting observation. An occasional summary of current hypotheses for a popular science readership is helpful.

The opposite process is called deductive logic, used for argumentation and proof in philosophy. A set of general statements (premises) is made, which are self-evidently true, or are assumptions. From these premises, a statement is derived about a specific situation. This statement will be true if the premises are true and the deductive logic sound. Deductive reasoning moves from the general to the specific. This linear process has been called top-down reasoning. For example, "Gravity makes apples fall to the ground. Apples fall to the ground only when they are ripe. Gravity acts only on ripe apples." This example shows a false conclusion, demonstrating why great care is needed to follow the philosophers' rules for deductive reasoning, which excludes much of a general population from this skill.

Enlightenment Science has successfully constrained the widespread use of false deductive reasoning. False deductions and unfounded assumptions had been widely used by manipulative people to trap the unwary into unwise decisions to part with their money or freedoms. However, the New Science has now revealed the limitations also of the Enlightenment Science method. It is effective only by controlling the physical ecology conditions to create comparable situations for measurement. As we have seen, life in the social and inner ecologies is an open system of informational repatterning, occurring by widespread communications that evolve. This is not a linear or controlled process. An open system is immeasurable in its qualities of changing relatedness. Since 2011, we now have to appreciate that the physical ecology is also an open informational system.

The truth criterion in an open system is transferability

The truth criterion in an open system for predictability is not that measurements are reproducible in different locations where physical conditions can be similarly controlled. When computer modelling, the truth of predictability is whether or not the predictions transfer into a different system, for example the mathematical system of a computer prediction into the physical system of what actually happens. When directly studied in an open system, such as a family, the truth criterion for predictability is whether or not a repatterning process *liberates exploratory and adaptable life over time within the observed system*.

There are two key phrases in that statement. Firstly, adaptable life means that the exploration in a living system such as a family is not merely individualistic and libertine or chaotic. Secondly, liberation does not mean breaking free from all relational constraints. On the contrary, systems predictably come alive by maintaining rich patterns of communication and by remaining responsive to widespread *feedback changes* (sensitivity). The truth is a deeply relational conversational and feedback process. Trying to control it for comparative measurement would disrupt or distort the system. However, the family, for example, can be intra-actively influenced by personal choice for *agreed action with others*. The ensuing feedback can then be compared with the anticipation. That feedback, especially if integrated from different sources, will be life-enhancing provided that *sensitivity to feedback of the impact* is wilfully maintained.

So, given the feedback systemic nature of life, is the notion of logic as defined by philosophers relevant to open systems? By definition, logic is the study of correct reasoning. It is about making inferences, guessing a conclusion or opinion about an outcome of a process, whether material or informational. Logic ensures that the process to reach that conclusion is informed by known facts or reasonably justified evidence or opinions. In general, consciousness, thought, and will are all clearly involved in making inferences logically. But we are looking here at a model of exploration of the unknown in which *personal values* (effectively anticipation patterns for life) are the main determining points to navigate by. Feedback cycles revolve around these anticipation patterns involving sensory analysis, correct orientation from the analyses, and integration into behaviour that is appropriate to the social settings that initiated the sensory inputs. In this cycle of exploration, anticipation patterns (personal life-values) have the same guiding role for choice of integrated behaviour as inferences do for purely mental reasoning. Deductive inference from premises to conclusion, and inductive inference from observations to hypothesis, both lead to behaviour that is justifiable as logical. So too in systems repatterning exploration, mapped patterns of *valued anticipations of a life-enhancing outcome* lead systemically to a logical choice of behaviour.

Mismatches from the anticipations initiate emotive reassessment and realignment of behaviour. In this way, emotion is completing the creative feedback loop. When 'guessing' one's way through life for sustainable living, emotion, when properly understood as systemic informational feedback about personal values, enables and empowers life-enhancing choices. It is important that people should not criticise themselves for their emotional experience. Recognising its vital role for healthy adjustment, people could rather become curious about it, and learn more about its informational meaning.

Nine features of a healthy, adaptive, dynamic system

To clear a way through to seeing the creative place of emotional curiosity, it will help to understand how interactive systems become learning and adaptive. Some key general concepts are transferable between all interactive systems at every scale of the cosmos. Examples are given below in the physical, social, and inner ecologies.

1. A system has several independent elements.

- Atoms

- People
- Inner organs as parts of body systems—which includes the brain as a nervous system organ that has itself many subsystems with interacting purposes.

2. The independent elements have rich and mutual interconnections.

- Waves and subatomic particles connect into the internal movements of atoms, which connect with other atoms into spinning molecules. There are simple rules of attraction and repulsion that shape the mutual interconnectedness of movement.
- People's distance senses detect sounds, movement, and physical presence (changing relatedness) so that they are sensitive to each other, for example in a family. There are simple rules of turning towards, or turning away from, that shape the movements that follow mutuality.
- Communication in the body happens by means of molecular hormones, neurotransmitters, cytokines in the immune system, and direct contacts between cell membranes in organs, through which molecules and electrolytes can pass directly, enabling coordinated change. The simple rules of physical attraction/repulsion are added to here by the simple rules of social attraction/repulsion, which add top-down hormone 'drifts' to the physical processes. The physical processes add bottom-up dynamics and limits to the whole inner ecology.

3. The independent elements feed back their changes to each other. The system as a whole coordinates into a 'dynamic state'.

- Water molecules in the atmosphere can coalesce into a cloud, which is a dynamic state of water in air.
- Conversations and body language in families can convey

subtle messages that create an 'atmosphere' in the family or household. If mutually constructive this might be called morale (light clouds); if mutually destructive this might be called abuse (storm clouds).
- Different subsystems feeding back within the brain can convert a state of heart-level awareness into an orientated state of consciousness (thought clouds).

4. Feedback also alters the inner states of the elements, which is memorised learning.

- Molecules can split, and can coalesce, bringing internal changes that linger after interactions.
- There are many memory systems among groups, such as spoken traditions or stories, which bring past attitudes into present-moment life, sometimes harmfully.
- There are many memory systems in the nervous system and among electrically sensitive cells in any organ system. For example, the adrenal glands grow larger if they are regularly overstimulated, trapping people in a vicious cycle of overreactions to change.

5. Learning alters the subsequent communications from each element.

- Atoms and molecules can be put into high-energy states that alter their behavioural response to interactions.
- Groups of people can absorb new knowledge that makes them react differently. In families, if one member learns to understand their anger in a more constructive way, the change of behaviour can spread its influence to others so that the whole household atmosphere changes.
- In the brain's subsystems, if a stimulus is pleasurable, then neural networks will alter their sensitivities to make it easier to seek and repeat that experience.

6. This relational process can establish a feedback pattern as a habit called an attractor state.

- Electronic devices can build up static charge that begins to affect functionality locally and in nearby electronic systems.
- Groups and families can develop habits of responsiveness to each other or to recurrences of stimuli from outside, for example neighbours. This may be called a 'family script'. Attractor states can be difficult to shift. It requires one or more individuals in the system to start exploring something new.
- Everyone knows what it is like to have a tune or thought going round and round in one's head for a period of time. Obsessional states are an extreme example of this same capacity for any nerve network to get stuck in a loop. A healthy balance is to have various habits of living that can also be modified, stretched, or broken when circumstances require responsiveness.

7. Subtle differences between each iterative cycle of feedback may cause fluctuations affecting the wider system.

- An internal combustion engine that is idling can start 'hunting' as its revolutions speed up, and then slow down.
- The productivity of an office or factory may vary at weekends and holiday seasons, but each weekend will be slightly different, and each annual vacation will have a slightly different effect.
- Every time a person writes their signature it is slightly different. The range is still recognisable as that person's signature, however. That range is the *fractal range* or the *fractal pattern* of that signature.

8. The habit can sometimes suddenly and unpredictably change into a different dynamic state, called a bifurcation of the

system, as a consequence of internal fluctuations and wider system influences.

- A dark cloud can suddenly burst into rain. It is a sudden change of state of the water molecules in air, where the wider ecological context has been changing.
- A human being in a family or workplace can suddenly burst into tears. The same principle of ecological change or relatedness patterns brings about this change of personal state in the social ecology. How the social system responds characterises that system.
- Anxiety-state chemicals released into the blood circulation may constrict arterioles, and prime blood-clotting chemicals to coalesce into clots, which may induce a heart attack or stroke. The whole individual has shifted into a different state as a consequence of stress.

9. A transitional time of turbulence may follow a bifurcation, until a new attractor state or steady state emerges among the feedback patterns between its internally modified elements.

- If water is flowing smoothly in a pipe (called laminar flow) and the pressure increases, it may start to gurgle in the pipe as turbulence sets in. Increasing the pressure some more may push the system into a new laminar flow state with a faster flow.
- If the government passes a new law, there may be a period of confusion as people have to adjust and relearn new habits that fit the new rules. Hopefully, a new set of simple rules will keep life moving comfortably, but turbulence may remain for some groups of people who are reluctant to adapt.
- When a person moves to a new town or country, they may initially take their old habits with them, but discover that some things clash or do not work well in the new culture. This may start a time of emotional turbulence,

with homesickness and regrets, until new relationships and new pleasurable experiences begin to accumulate, and new habits start to be formed. Then, on returning to the original home environment, another return culture shock can ensue with its own adjustment period, which will require a change of habits and quality of relationships with many people who have not had the same experience of a foreign culture.

Fractal scaling and replicating relatedness patterns

Responsive dynamic systems have intrinsic variability, such as writing one's signature, because every element and scale of the cosmos is in movement at some level. The chair you are sitting on is vibrant at its molecular level, and so are you. As different subsystems meet and interact, you and the chair, one mutually transforms the patterning of activity within the other to some variable extent.

This brings the notion of relatedness alive. The boundary between you and the chair is not as clear and distinct as your temporal lobe sensory association area would like it to appear in your informationally processing mind. To move your body, and to move the chair, require totally different action-planning strategies in your frontal lobes. They need to be distinguished as separate one from the other for this purpose, not seen as an integrally connecting feature of you intra-active as a personal substate sitting gravitationally in a living cosmos. Something at heart level will need to be excluded from the orientated frame of mind if you are to move that chair. The first thing that usually goes is the mindful awareness that we dwell with others in the domain of an open and ever-expanding gravitational cosmos that is our shared spiritual home.

Fuzzy boundaries between healthy systems

This dilemma is resolved by considering that, in dynamic systems thinking, all boundaries are fuzzy boundaries. The neocortex constructs them. The core brain bridges them, however, and the mindful person may have sensitivities that extend beyond them.

The patterning process within all forces and particles holds integrated movement into forms. The body and the chair are examples. Each patterned form has a measure of fractal variability at some level of its dynamics. At every boundary of every knowable form, the fractal patterns meet and mix. Forms overlap in their inner movements to a lesser or greater extent at their boundaries. Therefore, every boundary has itself a dimensionality of relatedness that is changing. Boundaries are alive in movement. A numinous light shines out from any meeting, full of potential, which is at least as real in its progression as the forms that we meet are real in their stability (temporary though most forms can be). Below the orientation horizon there is no distinct separation. There is movement transforming extending patterns. And each boundary condition has its own depth and breadth and length of presence to be experienced in its movements. Artists know this. Each boundary is a process that progresses its qualities of changing relatedness. To recognise this is to move from positivist philosophy to process philosophy.

This is relationality. On an extended scale some people call it the spirituality of life, while others flee from that intimacy into a preference for hard facts. Transcending forms reveals the potential for life to be renewed in the mutual encounters of people who see life differently and are willing to learn from each other.

Imagine two fields with a fence. Above the orientation horizon one person may have a mental habit that gives them pleasure when their temporal lobe has priority in its object recognition role. They may see two separate fields divided by the fence, in which they can do different things. Another person gets pleasure when seeing timeframes in which relationships can be seen to grow, which is a role to which the inferomedial frontal lobe sensory association area is recruited. This person may see the fence as the place where two fields meet. They may like the fact that there is minimum energy lost in transferring activity between them. A third person enjoys thinking about spatial arrangements generally, when his or her parietal lobe sensory association area is most active, and consults a map to work out how best to move between the two fields, simply acknowledging that that landscape is the way life is shaped.

Softening and hardening boundaries by personal values

Each of these three types of people will value different features of this rural scene. Their personal values will soften or harden the boundaries they see, matching the character of their chosen behaviour. Each may see in boundaries a numinous mental light of *potential*, which moves life from within their inner ecology. That initiated inner movement is part of their fractal range of variability as a living person, with their personality, spirituality, and character moving in their physical, social, and inner ecologies.

Relatedness is the overlap of fractal patterns at the fuzzy boundaries of objects. For most people, particularly with Westernised thought, object recognition screens out the potential for living change by hardening boundaries, reducing their potential for transformative relatedness, and thus veiling the light of potential in life.

As people recognise more clearly that the character of their choices can transcend the apparent boundaries of this world, their character becomes a transferable habit to any and every system that is potentially alive with movement. This means that fractal patterning can scale to higher levels of order, and can replicate that patterning of order in other systemic settings.

For example, rich networks of several elemental persons, all communicating subtly with many others, define the fuzzy boundaries of any social system that can potentially be brought to mind. Think about a family. How extensive is it? How intra-active are the people within it? Your sensitivities and thoughts contribute to defining a family system. Life-enhancing or dysfunctional choices impact this *adaptive system*. The system, however, already has various attractor states and habits learnt from its previous generations and its valuing of the future generations. If even one person in that family has had a habit of reacting to expressions of anger by hiding away, and learns instead that anger is informational about a person's hidden inner values, and learns how to enquire sensitively about what is important to that other with a view to making a different life-enhancing choice from hiding away, and speaks out instead to explore what can be done to name and then safeguard that value, then the behavioural expression of that personal value will change from frustrated anger to some other preparation, such as gratitude perhaps, or even an apology. And that change will fractally spread its influence throughout the whole family in time, affecting its future. It will also spread from there to affect workplaces, schools, and neighbours with whom family members relate through their daily, weekly, and longer boundaries of cyclical living.

Identical principles of organisation work among all these dynamic systems. A brilliant theoretical physicist called Mitchell J. Feigenbaum discovered a mathematical 'constant' that predicts the level of tension at which a bifurcation into turbulence is highly probable. The mathematics applies, for example, to a storm cloud discharging into lightning, and, I believe, to an unnamed value erupting into frustrated anger. Feigenbaum's Constant is one of those curious meeting points of pure maths with matter, and of matter with mind, and of mind with other people. Regulating life around the creative potential in boundary turbulence requires systemic logic to seed the re-emergence of creative order. From small beginnings, Butterfly Effect growth can fractally scale. But sustainable life requires a conscious balance to be sustained between all three principles of the organisation of movement—the changing relatedness of forms. For this, conversations with people who see life differently are life-enhancing.

Conversations

As previously mentioned, choice-making people are the independent elements of a social ecology. Choices arise from their inner states. Even at rest these states are stative, not static. They therefore have the potential to bifurcate under changing pressure into activity when the person's fuzzy boundaries overlap relationally with other people's fractal patterns and behaviour. Relationally, below the orientation horizon, personal values manifest as a fractal pattern of *emotional states* that prime responsiveness.

Therefore, when a conversation starts, the core informational content is the set of personal values below the orientation horizon revealed informationally as the person's fractal pattern of emotions. This set of values, activated by relevance to the changing situation, *is* personal identity. Unique kneejerk survival reactions arise from emotional fractal patterns, or behavioural traits of personality that are habits and therefore partially predictable. Also, reactions from emotional fractals can be moderated by reason or memories when that person's character of choice options is brought into play by extending their Big Picture of the situations they face. The core informational content of a conversation is *not* the words spoken from the left cortical hemisphere. Spoken words ride on the emotive waves of relational movement. Words have unique relational frames of meaning for individuals. They are given their boundaries in the neocortex and then strung into a linear syntax, a process that converts a global pattern of neural activity into a linear string that people hope is effectively informational. Linear restructuring occurs above the orientation horizon. In this informational realm most people assume that reality is objects in space-time, and that the conversation is about these objects. But the core conversation is deeper. It is more moving and extensive than that. People in a conversation *are* their priming values from their unique perspective into the informational cosmos. And values do evolve and mature as life experience grows from the many seeds that are planted in people's inner hearts.

Vast amounts of subtle messages are conveyed by emotions conversationally into and through the fuzzy relational boundaries with other people. For life to be based sustainably on human values, not on linear robotic data exchange, sensitivity to heart-level resonance at the mutuality of boundaries is vital. Mirroring facial expressions and body postures is part of the emotional intelligence that generates this internal reflection. Unfortunately, that term 'emotional intelligence', abbreviated to EQ, has been twisted in motivational and management business systems. In those settings, an Old Science belief in the measurability of EQ as a feature of personality has led to tick-box self-assessments. These cannot capture the dynamic mutuality of emotional intelligence, however.[2] I shall avoid using that confusing term, instead using Emotional Logic to convey the notion of healthy adaptability as previously described. Understanding the *emotional logic* of a healthy adjustment process, and being curious about its mutuality during times of change, builds character formation by strengthening responsive relatedness.

Sensitive dependence of conversations on initial conditions

One final point about conversations is important to make. Movement may be considered mentally to start from an initial set of conditions. Repatterning transforms these conditions *within the timeframe that has caught someone's attention.* We have previously considered how, within any system, slight variations are continuously on the move. The science of Complex Adaptive Systems (CAS) has found that those slight variations have a profound and magnifying impact on the way the system movement evolves. This means that 'initial conditions' will also greatly impact the way conversations develop. It is called *sensitive dependence on initial conditions.*

In the social ecology, for example, mood can affect the way a theoretical subject is discussed among a group of people. Clearly, the Stoic and rationalist response would be to work hard to regulate or exclude mood and emotion from the discussion group. However, that creates a closed system for only data transfer. Given the overall purpose of a group, that may be exactly what is needed; but it is not a rule that can be generalised into a strategy to manage the open system of sustainable social life as a whole. Small changes in open, non-linear systems may have big effects. This is captured in the proverbial poem, 'For Want of a Nail':

For want of a nail the shoe was lost.
For want of a shoe the horse was lost.
For want of a horse the rider was lost.
For want of a rider the message was lost.
For want of a message the battle was lost.
For want of a battle the kingdom was lost.
And all for the want of a horseshoe nail.

Most people have heard of this as a Butterfly Effect without knowing where that term came from or what it implies. Among the early chaos theorists was Edward Lorenz, a mathematician working on weather prediction in the early 1970s, who noticed that the path taken by tornadoes was greatly influenced by small perturbations of non-linear dynamic factors in the environment.

He described this poetically, saying that a butterfly flapping its wings even before the tornado had formed could alter the path of the tornado.

Butterfly Effects are therefore not the same as ripple effects. Butterfly Effects are the fractal scaling of dynamic repatterning through more extensive overlapping systems at their fuzzy boundaries. This repatterning is non-linear. It is *deterministic,* meaning that relational causes have their effects, but the *exact* outcome of patterns and timings is *unpredictable.* General trends in repatterning will be identifiable, but the process is probabilistic.

A ripple effect, on the other hand, is the predictable spread of a cyclical change more widely through a single system. To compare the two we could return to the football stadium, where a Mexican Wave is a ripple effect through the crowd occurring because there is a consistent heart-level agreement between the elemental persons to cooperate. A Butterfly Effect would be when someone in the crowd shouts out that there is a fire outside their section. Movement may then spread by curiosity through that section, and eventually to the whole crowd and the competing teams, but with no overall agreement on how to respond. Many may respond with an orderly exit, some will decide to stay and see what happens, and some will panic and cause chaos in their locality. All this behaviour is part of the fractal range of scaling of the call, "There's a fire outside".

A conversation can be viewed as a series of Butterfly Effects between two or more people. The *initial* conditions for a conversation start before the first words are spoken. However, these initial conditions do not necessarily determine the course of the entire conversation from there. In fact, they seldom do. There is also no clear end to the timeframe of a conversation. So, how far back are the initial conditions for life upon which our present conversational experience is dependent? The universe does not work like clockwork, so any initial conditions that set it moving do not determine everything that has happened ever since. The universe is an open conversational system, a cosmos in which life is inherently in eternal potential for growth and diversification through subtle small variations of feedback. The safest way for human beings to think about initial conversational conditions in an open cosmos, from quantum to gravitational scales, is to start from whenever a human being makes a choice. Where options exist between different possibilities at a boundary, a new set of initial conditions for movement is relationally created. In this way, conversational conditions for the unexpected to develop into fractal scaling are ever-renewed.

Personal values are physical settings in core brain filters

If the initial conditions for a conversation include choice, how does a conversational adaptive process become stabilised?

In Part I we saw that anticipation patterns for the impact of behaviour are constructed in the frontal lobes of the neocortex and, on their way out to the musculature, are also embedded in filters to subsequent sensory feedback in the core brain, as synapse-on-synapse connections. Emotions are generated by mismatches, or resonances, between these filters and environmental feedback. These limbic core brain filters thus *physically embody* the memories and anticipations that *are* human values for survival and thriving—not necessarily in great detail or specificity, but as probabilistic trends. Values are not only psychological phenomena, or linguistic concepts. Personal values physically are informational filters to social feedback located in the core limbic brain.

The emotive fractal with which these personal values manifest is the core of what it means to be a person.

The award-winning work of neuro-clinician Antonio Damasio has been central to a revolutionary shift of thinking within the scientific community about the informational value of emotion. Starting with his famous *Descartes' Error: Emotion, Reason and the Human Brain,*[3] a subsequent series of thought-provoking books trace his and his wife's research programme.[4] He builds a picture of emotion as preverbal *'sensory markers'* of changes of the inner physiology. These sensory markers enter the brain's neocortex via an area of cortex hidden deep in the sides of the hemispheres, called the insula. Processing here contributes to the sensory association processes to orientate the person with a mental *body image*. This image also locates feelings of emotion and activates memories to give meaning to them.

In the triquetral understanding of orientation being presented here, this body image of 'self' may become objectified in the orientation process. It may then mistakenly be understood as if it is a psychological self, perhaps a self-existent entity (a homunculus, a soul, a spirit body) that lives parallel to the physical body.[5] Correctly understood in a triquetral framework, body image is the informational summary of our *changing relatedness* in physical life. The triquetral framework allows (but does not require) relatedness to be thought of as spirituality. Qualities of relatedness can be both local and distant and Big Picture. A heart-level awareness of distant resonances may be brought into conversational consciousness, where they will seem mystical because they do not fit within the orientation in space, time, and present experience of the changing local qualities of relatedness. How people mentally 'handle' this incongruence is part of the infinite diversity of human endeavour to make sense of life, and to stabilise or predict what is likely to happen next so that they can emotively prepare.

Integrating Damasio's work on emotion into a triquetral orientation framework, the experience of emotion during changing relatedness will be included into both the body image and into a person's spirituality of connection or disconnection in life. It is foundational to an individual's developing and evolving or maturing sense of their personal identity.

Damasio found that if physical emotion is missing from the body image, people cannot choose between two or more options that are *rationally equally valid,* such as choosing a diary date for a follow-up appointment. Emotion, values, anticipation, thought, and will are all brought together when making a choice of behaviour. Emotions 'weight' the decision with values. Distortions of body image will affect the quality of resulting choices. These distortions are a neocortical influence into the core brain sensory filters. Therefore, the neocortex does not only generate pure reason and a rationally virtuous or moral society. Absence of emotion can disable the power of choice between options of behaviour. Excessive emotion included in the body image may overwhelm or distort conscious, named access to the key values that might otherwise have empowered a maturing decision.

The Stoics aimed to break this emotive relationship away from 'mind' in order to prioritise mental reasoning. The noble purpose of *introducing* reasoning into social dynamics can be generalised too far, however. In our technologically advancing and materialistic world, many people have become socially isolated through not being taught how their emotions are clues to building a future based on people's values. Nevertheless, the New Science can restore hope by adding good reasons to broaden the Stoic strategy for making will-based decisions. In the next chapter, a practical method is described that maps emotional patterns and transforms people's understanding of their helpfully informational social role. Thought, will, and emotion can be restored in consciousness to a healthy partnership for action in the world.

Chapter 5

Emotional Chaos Theory and Emergent Order

Adaptability is systemic repatterning of values

Imagine you are walking along a street and someone looks at you in a strange way, worryingly, perhaps threateningly, it seems. In an instant your emotional chemistry can change. Emotion is rapidly-changing *social physiology*. Mood is slowly-changing social physiology. Atmosphere is social physiology that you walk into.

Mood is a variable part of the initial conditions of a conversation. Seldom is it taken as a stimulus to name the personal values that are lingering in the core brain's filters. Those values influence the feedback cycles for learning, exploration, and the power of choice, however. The model of healthy adaptability to be presented here is based on developing the skill of naming those filtering values given insights from the New Science of adaptive systems. The strategy then moves on to choosing actions that address probably just one of those values rather than just trying to regulate or control mood or emotion. The word 'affect' (a noun) becomes important in this New Science perspective on adaptability in living systems. Even more generalised than mood, affect is the pervading background *disposition* of mind or heart to emotionally bias thought, action, and reaction. The language of colours is commonly used to describe affect and mood. Affect is mostly relevant to a person's variable capacity for movement or stuckness, in the sense of releasing or trapping personal energy to face changing situations and relationships. In UK English, affect relates largely to 'being bothered', or having attitude.

In the early 2000s, a systemic method to map the emotional patterning behind mood, affect, and emotional reactions was developed that is now being taught and used internationally, called Emotional Logic. It is a paradigm shift in understanding how unpleasant emotions fit together to make an integrated and healthy adjustment process in response to change. The Logic is the healthy process of adjusting during setbacks, disappointments, or hurts. Emotion is understood as personal energy getting organised to achieve those adjustments. E-motion is energy in motion. The way adjustments in lived experience do *not* proceed logically can be mapped using simple tools to show the emotional patterns that can trap personal energy inside, leading to a sense of stuckness in life, perhaps with distress, tension, or confusion. Trapped adjustment energy obstructs personal responsiveness or 'flow', and hinders responsiveness in relationships. By reflectively sorting specialised 'emotion cards' into patterns, and by ticking relevant emotion spaces on worksheets, the *cerebral* naming and rationalising of emotion is short-circuited. Emotional Logic thus uses kinaesthetic learning methods, opening the neurology of change and learning by focusing on the sensations and hand movements while recalling memories. This maps core limbic brain, preverbal, complex preparation states for movement.

The need to string words into linear sentences evaporates as insights are gained from the spatial patterning. Conversations start easily from teaching on the backs of the Emotional Logic cards, which brings a sense of safety to talk about in-depth and sensitive subjects. Confidence and relational trust can grow proportionately. The mapping method opens a new lifelong learning approach to personal development by releasing energy to explore new possibilities in life shared with others. Reconnecting inner heart and rationalising mind then happens by solution-focused action-planning around a chosen named personal value. This opens a growth attitude to the previously overwhelming situation, because the personal value can now be talked about with others, and agreed action anticipated in response to the situation being considered.

This is a learning strategy for personal development of new capacities to prevent problems from developing. It is not a therapy, because it has global application in its *learned honesty* approach to any new situation. Learning to activate one's genetically inbuilt Emotional Logic does not work simply to relieve symptoms or improve a particular function. It develops core personal identity to adaptively respond to changing relationship patterns by improved powers of choice. Problems do get resolved along the way, but the greatest gain is in strengthening relatedness and reducing isolation. By learning to informationally map an inner systemic preparation state for action, personal identity can grow to add choices. Then, behaviour is no longer just reactive to felt need. It becomes a deeper informational patterning of the whole inner ecology, including *affect* emergent from personal values. When a person knows that this affect means that they do have values, even though they may never have named them, they can harness their affective energy into rationalised, values-based action. This prevents the common mental illnesses, distress, and socially disruptive behaviour that may otherwise emerge when accumulating loss emotions clog the feedback loops into core brain filters. The solution is not to get rid of emotion. The solution is to understand the values that are the roots of that reactive energy, and to turn that energy into inner strength for healthy adjustment.

Overcoming the practical difficulty of naming personal values

People commonly have difficulty naming their personal values, however. If I were to ask, "What are your values?" I would be almost inevitably greeted with a bemused silence, and then a hesitant, "Well, that's a strange thing to ask. I suppose food, family, a bit of fun at weekends..." Alternatively, some may revert to what I call the heroic values that they have been taught is the right answer by schools, business organisations, or religious teachers, such as honesty, integrity, transparency, perseverance, courage, and so on. These are all *aspirational values*, not *personal values*. If, for example, someone is being physically abused by an alcoholic relative, the personal values may be completely in conflict with these aspirations. They may include: concealment of bruises, a hiding place, avoidance, safety, a friend. If I am living in physical poverty, I may still try to develop the noble character of aspirational values as I explore how to survive, but oppressive or dangerous relationships make that difficult. In general, aspirational values cannot be lived as an individual until some personal values that are influenced by relationships are secured.

Personal values are difficult to name because they dwell in the domain of life below the orientation horizon. Heroic, aspirational values dwell above that horizon, informationally part of a constructed or even mythical inner world of ideas. In practical, everyday living, people only know what they value when they see a risk that they might lose it, or have lost it. And they only know that something important has been lost when the unpleasant loss emotions have been activated from the preverbal core limbic brain. The loss emotions are not 'negative'. They are vital information about personal values being challenged. Usually the emotional reaction is initiated before the cerebral cortex has recognised or named what the issue is.[1] The emotions are informationally there to initiate a search for a cause cerebrally. Merely activating a comfort-seeking desire to get rid of the emotions means that people will never engage with their values. They may drift through life instead as a passive victim of their emotions. Connecting heart and mind around naming those values, however, empowers the inner ecology to contribute to shaping the social and physical ecologies, 'earthing' our humanity. Thinking more deeply through the emotional experience to recognise the personal values that are being challenged empowers people to become truly realistic in their hopes and plans in the way they relate to others and shape life.

Therefore, the difficulty with asking people simply to name what they value is overcome by first legitimising their loss experiences as evidence of their humanity. Understanding unpleasant emotion as a healthy survival adjustment process restores self-respect sufficiently to consider naming their losses and writing them as a list. The list of hidden losses that has been driving their emotional experience then can be converted (we say 'turned upside down') into a list of their personal values, previously unnamed. The loss list becomes a personal identity list instead. It shows who they are. It shows why they kneejerk react in the world, their personality, affirming their value as a human being because they have values that were previously unrecognised.

Below the orientation horizon is the realm of awareness. Kinaesthetically mapping the moving patterns of emotions there simultaneously maps how values are shaping the inner heart's responsiveness to the world. By mapping this externally using cards and loss reaction worksheets, empathic others can see into each other's inner hearts in a new way, bringing heart-level values and habits of response into shared consciousness. In this way, features of complex patterns can be named, compared, understood, and discussed without feeling that the patterns need to be converted into a linear logic, or translated into linear sentences of words. People feel heard when conversations shift from talk about feelings and behaviour to personal values and agreeing action plans.

Mapping emotional chaos

Figure 12 is a screenshot from an online 'card pattern generator' that enables people anywhere in the world to show and share safely how they feel about a situation they face, or one that in the past has led to setback or disappointment.[2] We encourage people by saying, "Don't *tell* me how you feel. Use these cards to *show* me safely how you feel." The pattern shaped manually is infinitely more informational than could be summarised in one linear spoken sentence of symbolic language. This example was laid out by someone who had an obsessional compulsive drive to self-harm when facing apparently insoluble problems. It shows the complex patterning of their inner heart dynamics when feeling the drive. The behavioural reaction comes out of the whole pattern, not from any one emotional component.

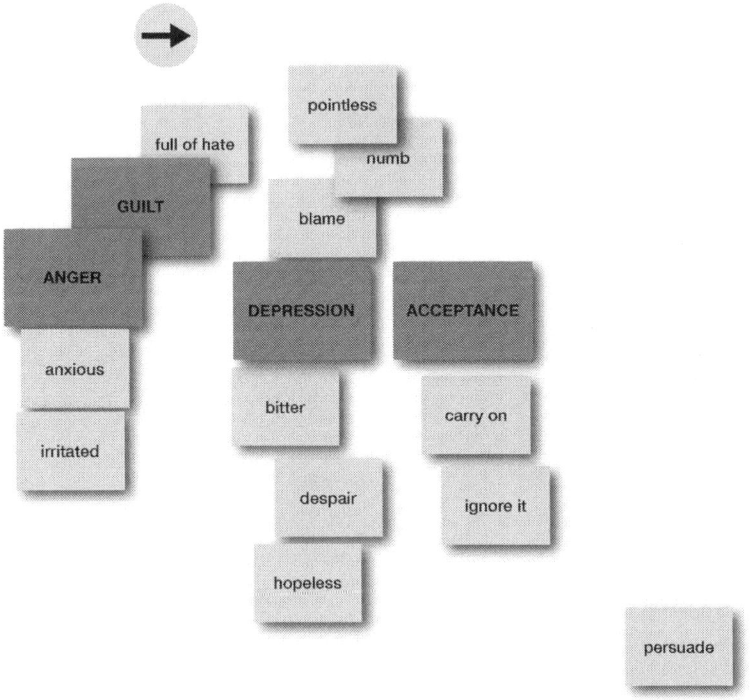

Figure 12: A pattern of loss emotion cards from the Emotional Logic Centre 'tools site'

Seven orange cards represent the core emotional adjustment states researched extensively in the British hospice movement and elsewhere—Shock, Denial, Anger, Guilty self-questioning about one's own role or abilities, Bargaining when yearning to recover something important, Depressive feelings of emptiness and powerlessness on seeing one's limits, and Acceptance of the loss of something valued with letting go while also being able to explore new beginnings. We call these inner states *emotional Stepping Stones for adjustment*, because they are firm inner places of self-organisation on which to get balanced and from which to move on to another emotional state as the adjustment process progresses.[3]

Emotional Logic widens the relevance of loss-related emotions (grieving) from bereavement to any valued aspect of life, and shows how multiple small losses in daily living can accumulate into a complex state somewhat like bereavement. These seven emotional 'preparation states to respond' are not stages in a process. The process can become chaotic and not sequential. People jump back and forth between the states as if each were a stepping stone being used to cross a shallow river or stream. People may miss some out, choose not to go where they look slippery, or get stuck straddling two of them while aiming to get to a more secure way of living on the other side.

On or in each of the emotional preparation states for movement, an individual may have a far wider range of *feelings of emotion* as the preparation and situation engages with their memories. These feelings fill out the corona of lived experience. The memories can feed back into anticipation patterns to add to the complexity and unpredictability filling the inner heart. Twenty-eight pale green cards (a complementary colour to orange) represent a selected range of feelings of emotion that individuals may describe in their relevant communities and languages. A few of them were used in Figure 12. Laying out card patterns slows down emotional processing and reactivity.

Reasoning and learning can then fill a creative conversation to restore a healthy partnership between mind and inner heart, and to restore constructive responsiveness and curiosity between people who see life differently.

Recognising harmful effects from whirlpools of loss emotions

The patterns laid out are infinitely variable. Some show inner emotional chaos and overload, others show inner order, and others show restrictive oversimplicity. In Figure 12 the pattern shows a feature of emotional complexity that can only be detected by the Emotional Logic tools. The client placed the Anger and Guilt cards overlapping. This pattern we call a 'whirlpool of loss emotions', because it suggests that when the person experiences anger, they also feel guilty, and when feeling guilty they also feel angry, perhaps so much that the body chemistry cannot keep up with the inner heart and so a physiological chaos results. Dissociation of feelings from thoughts and an accumulation of distress, tension, or confusion will follow.

With thousands of cases across five continents, we are confident that this particular physiological turmoil and its emotional entanglement is associated predictably with obsessive compulsive drives, or perfectionism, which we have found can spontaneously resolve on understanding how these two emotions each have different useful purposes in a healthy adjustment process. This systemic link, between understanding the healthy adjustment process and the choice to let heart and mind cooperate in harnessing that emotional energy with a rational purpose, prevents the formation of whirlpools. This principle at work in various entanglements among the seven healthy emotional Stepping Stones for adjustment has been shown to prevent the narrowing of responsiveness into habitual self-harming behaviours and common mental illnesses such as anxiety and clinical depression. Solution-focused order can then emerge instead out of that emotional chaos following a psycho-social learning process that builds upon named personal values.[4]

Emotional Logic coaches learn how to interpret these patterns, and to agree *learning plans* that will untangle the features and prevent further trapping of personal energy. Coaching conversations revolve around understanding stuckness and restoring movement, releasing the energy to explore new beginnings with new directions and new depths of relatedness.

It is not the purpose of this book to explain the healthy adjustment process in detail. This can be learnt from resources available from the Emotional Logic Centre.[5] The purpose here is to explore the fuzzy boundary between the unique personal inscape and the external landscape of life's movements, the liminal zone of intra-action. That fuzzy zone of choice-laden intra-action is the factual reality of relatedness. In this liminal zone, shown iconically where two personal triquetrae meet, the physical, social, and inner ecologies all overlap and mutually transfigure. This liminal repatterning is where healthy adaptability is birthed. It grows *within* an individual person's lived experience. However, the renewing of life is being continuously reborn *in the open system of their relatedness*. When an individual becomes a 'closed system', individualism turns a person metaphorically into a snooker ball bouncing off others until entropy (or prison, or illness, or social isolation) cools it all to a bland halt. The opposite opening process is to see mutual responsiveness grow as the warmth of kindness or mercy in aware and sensitive relational connections. This may sound like a utopian dream, but there is a hard reality and truth to it as the path to renewal after traumatisation, and many people who are comfortable now will feel traumatised when their environments crumble.

Mapping the personal inscape

Emotional Logic describes a 'life cycle' of movement, shown in Figure 13, revolving around personal values.

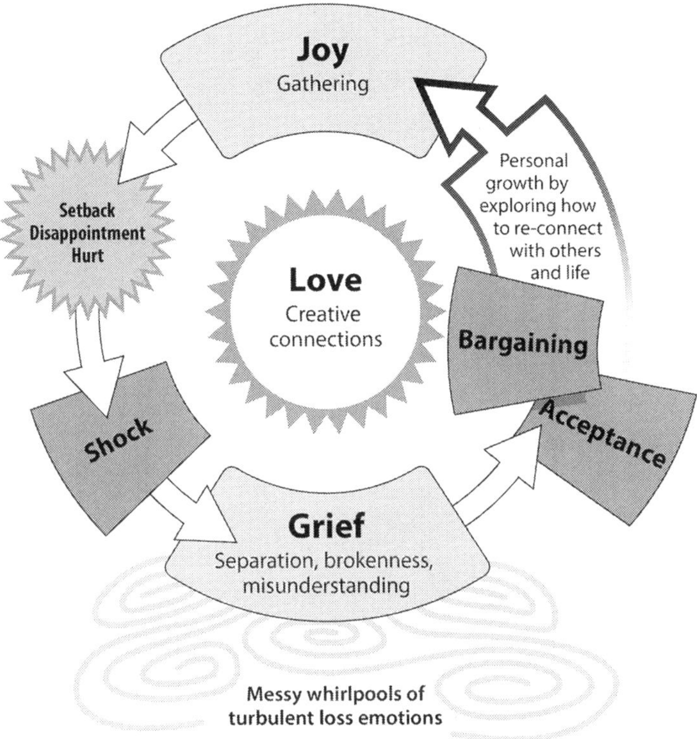

Figure 13: The Life Cycle diagram—movement revolving around personal values

Personal values are 'what this person loves', in the sense of an enduring creative connection at heart level that forms part of the person's identity. A person is moved either towards or away from situations on the basis of their personal values. The point being made in this diagram is that love has two forms, a stative joy when things are going well, and an active grief when separation, brokenness, or misunderstanding threaten the bond. People who devalue emotion have not realised that these unpleasant grief emotions are not negative. They have the positive purpose of moving people to seek a personal growth cycle around preserving their named values during times of change. Loss emotions are not the end of love. Grieving may at first feel crippling, but it moves the person's inscape repatterning constructively in time towards the top right-hand corner of this diagram, where the energy is released to explore new ways to reconnect with valued people, things, and interests. The purpose of this movement and repatterning is to restore love to its joy mode, where life as an open system becomes creatively diversifying, stabilising in its mutually respectful diversity, and potentially self-replicating.

To name the hidden losses underlying patterns of loss emotions, we teach Emotional Logic coaches and skilled helpers to use guess-checking methods with people who feel stuck. Figure 14 shows a short 'hidden loss list' gathered during a conversation with someone who had experienced an interview for promotion that went embarrassingly wrong.

Asking what was important about the situation that will be missed is a way to talk about loss safely. The conversation can be guided to explore wider areas of life than the immediate experience. For example, any issues with work colleagues, life partners, and in any other interest area of life will add a background, multi-systems liminal zone of relatedness that transcends the immediate locality of the situation. The issue in this interview example is not managing the stress of the interview for the individual, but managing its wider impact in the future. Having named the hidden losses non-judgmentally, the conversation becomes creative when the skilled helper emphasises that this is in fact a list of the person's values, each of which is starting its own loss reaction. By asking the person to reflectively tick which loss emotions arise when thinking in turn about each value, a pattern of emotional ticks emerges. This maps the inner tension that could cause a kneejerk response whenever this situation is remembered.

Named losses	Shock	Denial	Anger	Guilt	Barg'n	Depr'n	Accept
The promotion	✓	✓	✓	✓		✓	
Face - at interview	✓			✓		✓	
self-respect			✓	✓		✓	
joking with colleagues				✓		✓	
hopes for more money	✓	✓	✓				
holiday with my partner	✓			✓			
can't get anything right	✓		✓			✓	
name a loss							
Totals	**5**	**2**	**4**	**5**	**0**	**5**	**0**

Figure 14: An Emotional Logic 'loss reaction worksheet'

The patterns mapped like this have been proved to be predictive of behaviour. For example, we could predict that the pattern of unrecognised grieving shown in Figure 15 would give the person who failed the interview a fatigue state, while the grief pattern in Figure 16 would prime the person to be confrontational or destructive, or perhaps to have self-destructive suicidal thoughts.

These tick patterns on the loss reaction worksheet map the interruption to the flow of emotional adjustment energy emerging from personal values. Figures 15 and 16 show this flow turning around on itself into a shallow whirlpool or eddy as if turbulence occurs in a stream between two of the emotional Stepping Stones, or in a whirlwind of dust over a landscape. When two different personal preparations to respond magnify each other, or conflict, the biochemistry and biophysics of the inner ecology gets pushed into a similar turbulence if overloaded, or even into chaos, which will affect social messaging and internal messaging to the cerebral cortex. The person then finds it difficult to make sense of their own feelings. Other people also do not know how to interpret their behaviour and their spoken statements of distress, tension, or confusion. Common mental illness diagnoses result, and socially disruptive behaviour, most of which emerges in this way out of unrecognised grieving that has accumulated over long periods of time without the insight needed to transfigure it. We have found that giving people this insight through sharing the Emotional Logic mapping of emotions and values opens the door to creative conversational feedback. It is that feedback, called knowledge, that transforms and spontaneously releases adjustment energy from the whirlpools or whirlwinds that have been disturbing and trapping people's inner hearts. Healing and renewal of life and hope follow, sometimes with remarkable ease and joy.

Named losses	Shock	Denial	Anger	Guilt	Barg'n	Depr'n	Accept
The promotion	✓			✓		✓	
Face - at interview	✓					✓	
self-respect	✓	✓				✓	
joking with colleagues	✓					✓	
hopes for more money	✓					✓	
holiday with my partner	✓					✓	
can't get anything right	✓					✓	✓
name a loss							
Totals	**7**	**1**	**0**	**1**	**0**	**7**	**1**

Figure 15: A Shock-Depression whirlpool producing fatigue

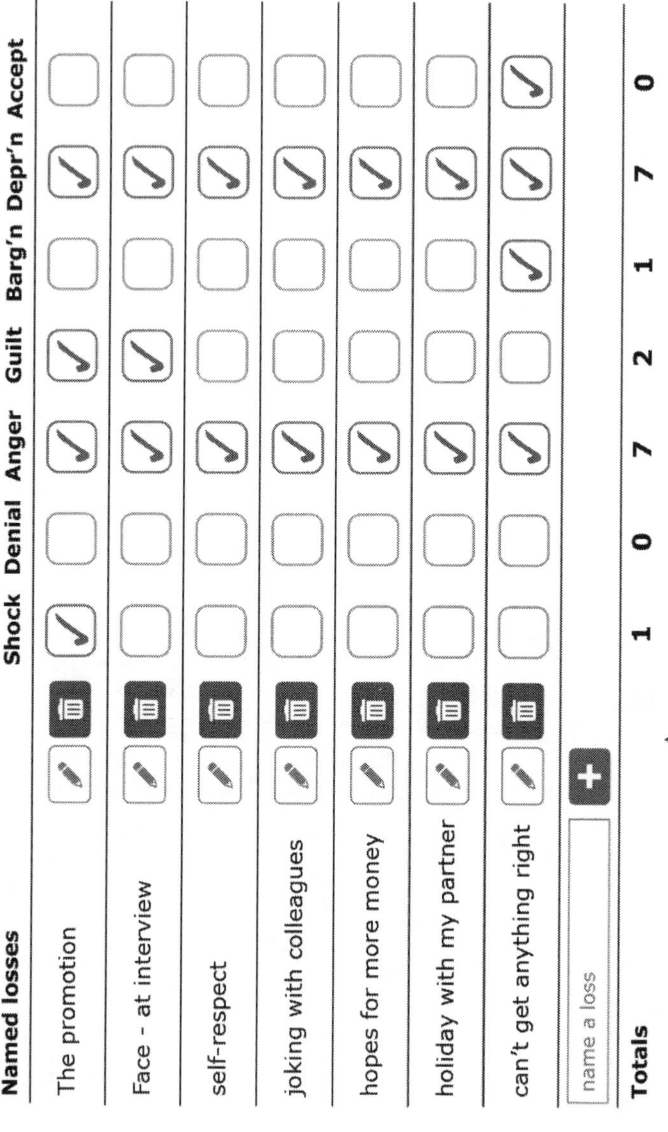

Named losses	Shock	Denial	Anger	Guilt	Barg'n	Depr'n	Accept
The promotion	✓		✓	✓		✓	
Face - at interview			✓	✓		✓	
self-respect			✓			✓	
joking with colleagues			✓			✓	
hopes for more money			✓			✓	
holiday with my partner			✓			✓	
can't get anything right			✓		✓	✓	✓
name a loss							
Totals	**1**	**0**	**7**	**2**	**1**	**7**	**0**

Figure 16: An Anger-Depression whirlpool producing destructive drives

In the loss reaction worksheet, we are able to see both the list of values that *is* a person's living identity in the world, and their coronas of emotional physiology, which messages those core values. The loss reaction worksheet tick pattern is a unique reconnaissance photograph, or X-ray, of the inner human heart, showing a snapshot of its dynamic patterns reshaping the liminal zones of relatedness within, and as importantly beyond, the self.

The inner ecology as a complex adaptive system

The inner human heart is a complex adaptive system spanning all three ecologies. Here, synergic flows of informational energy within filtering neural networks in the core brain can become habitually stuck in maladaptive attractor states. These become mini-closed systems, insensitive to energetic changes. These attractor states narrow the range and power of choices that people can make. Whirlpools of loss emotions *simplify* complex or chaotic emotional states, but maladaptively. They trap people in reduced inner worlds. The associated distress, tension, or confusion behaviours fractally magnify the impact of this entrapment in shared life-settings, transforming the environment to some extent.

Enduring connection is shaped and repatterned in movement by personal values. Personal values are themselves reshaped and repatterned by feedback between the core brain and neocortex. Reshaping personal values is personal development, because the inner human heart is the core of being a relational, conversational person. Love's grief mode for accumulating small and large losses can turn to exploratory joy on reassessing the corona of loss emotions and the values behind them, with maladaptive behaviour spontaneously resolving over time. To achieve this, it will be helpful to see how the two cerebral hemispheres may contribute different types of informational feedback to the core brain's balancing, embedded as it is within the wider social ecology. The inner human heart holds adaptive paradox in balance. It is not merely robotic.

It may be helpful to review here the nine features of dynamic systems applied specifically to the way the core brain can balance choices. Different input from the two cerebral hemispheres to this complex adaptive system is considered after that.

1. A system has several independent elements.

Each personal value is an independent informational element of the personal identity. These personal values are anticipation patterns. They are embedded in the core brain's synapses to filter incoming sensory data. Therefore, they are themselves dynamically changing as anticipations change. These priming patterns in the core brain are like a word that has been framed in the neocortex—but they are unspoken as verbal language. Several values can be held overlapping in the filters. They feed back to the cortical levels in an intermingled, complex way via preverbal language, physiological mood, affect, and emotional reactions, which appear in consciousness as different types of feelings.

2. The independent elements have rich and mutual interconnections.

The way neural networks work includes 'surround inhibition' and enhancing recruitment, both simple rules of networking. These arise through rich mutual interconnections between neurons, so that strong incoming informational data can be amplified and passed on with priority over relatively less interesting background changes. The Life Cycle diagram and other Emotional Logic materials summarise the simple rules of healthy emotional adjustment from one state to another within the wider social ecology, as that impacts the core brain preparations for responsiveness.

3. The independent elements feed back their changes to each other. The system as a whole coordinates into a 'dynamic state'.
Functional MRI scanning tends to show the whole core brain being active, and the neocortex being selectively active. These states all show the richness of feedback loops that bring the whole brain, ideally, into a coordinated, resonant state of activity that is informationally a dual aspect to being consciously responsive in the wider physical and social ecologies. That ideal state of whole-brain involvement is probably a myth, however. Consciousness relies on comparison of differences between features of awareness. Comparison and weighting of options generates progressive inner dynamic states, while others evaporate.

4. Feedback also alters the inner states of the elements, which is memorised learning.
Feedback loops from the core brain to the neocortex pass via thalamo-cortical and other neural routes. In the neocortex, the left and right cerebral hemispheres have different patterns of processing before feeding back down, where they simultaneously activate the body and revise the anticipation values in the core brain. Learning occurs at every level of this feedback system. Synapses grow larger, more numerous, and carry larger packages of neurotransmitters where they are heavily used. Less well researched is the potential role of biophysics in memory. Every biochemical change is also a quantum field change of probabilities, like a halo influencing nearby qualities of relatedness in liminal zones. These liminal field movements may add to the evolution of spatial patterns in biochemical reactions, which would also constitute a feature of memory.

5. Learning alters the subsequent communications from each element.
Living neural networks alter their physical structures with activity, unlike computer discs. Neural networks are all residual

databanks, throughout the body and its brain. These databanks are informationally reactivated by an approaching change in their new formation. Their shape and structured form therefore influence the progression of movement at every systemic level of life.

6. This relational process can establish a feedback pattern as a habit called an attractor state.

An embodied person is intra-active within wider physical and social ecologies. People develop habits and cyclical activity patterns. The important feature for adaptive life is how resistant or responsive these habits are to variation, when informational movements approach from 'outside' the learned habit. If the habits have become a closed system, the lack of responsiveness will have a deathly effect on the surrounding ecology, and the person's life with that habit will tend towards entropic blandness. A balance needs to be found between healthy habitual cycles and healthy variation in response to change with an ability to explore new beginnings.

7. Subtle differences between each iterative cycle of feedback may cause fluctuations affecting the wider system.

In the human being, subtle differences in habitual cycles are a sign of the capacity for healthy responsiveness. For example, the pulse rate is constantly varying very slightly as background preparation states informationally change. When a heart tracing, an ECG, shows absolute regularity of the pulse without beat-to-beat variation, it is taken as a sign of heart disease. The heart is functioning then as a closed system, not matching its activity to the wider needs of the socialised body. The same principle holds true for cortico-core feedback loops that embed values in the core brain filters. Subtle variations there will prime the person for responsiveness, which means that new changes can be fractally magnified through the whole person.

8. The habit can sometimes suddenly and unpredictably change into a different dynamic state, called a bifurcation of the system, as a consequence of internal fluctuations and wider system influences.

The internal electrical conduction system of the heart can become faulty, tipping the state of heart muscles into dangerous and life-threatening arrhythmias. A healthy contrast is, for example, how sportspeople experience the 'second wind' effect, when their physiology moves through a critical point from resting mode to dynamic responsiveness mode. In the core brain, the equivalent is waking up, or becoming alert to some new change. The former mode of activity has bifurcated from some previous stative or habitual state into a different dynamic system state.

9. A transitional time of turbulence may follow a bifurcation, until a new attractor state or steady state emerges among the feedback patterns between its internally modified elements.

Conflicting feedback from analysis of options in the neocortex, which is constantly priming the core brain filters, may lead to behavioural hesitation. When a balance of values is found that justifies a risk of embarking on new behaviour with an unpredictable outcome, the *choice process* is coming to its critical point of decision. At this point the core brain filters bifurcate into a new state. This state will preserve some older values and include preparing new alerts dependent on other situational values. For most people, change and decision-making is an emotional process. If that emotion is not understood as vital information, or the values remain unrecognised, turbulence may accompany the inner state's bifurcation, either into uncertain action, or into resistance to action and withdrawal from situations into an inner ecology turmoil. If the life cycle of emotions is understood well enough to bring heart and mind into a living systemic partnership, then a chosen and mindful course of action can be pursued with a full inner heart, not constantly doubting the value of that course. Qualities of relationship will then progressively change towards a new stability of inner and social levels of ecological order.

Feedback from the neocortex into the inner human heart

Early neuroscientists assumed that the two cerebral hemispheres housed the human person. It seemed obvious that these hugely overgrown parts of the brain compared with other mammals in an evolutionary tree would be the driving seat, or armchair, for the assumed 'homunculus' of a thinking, rational person. However, experimental and therapeutic brain-scanning work from the turn of this third millennium suggests otherwise. The two hemispheres have importantly different functions. The left and right hemispheres are indeed richly connected with each other across the midline in the corpus callosum by nerve fibres that bootlace back and forth. However, these fibre connections are surprisingly *not* vital for a sense of unique personhood to emerge (in dual aspect) from neural loop activity.

A strange discovery was made by clinical neurosurgeons who were attempting to help the lives of some unfortunate people who were severely troubled by epileptic seizures. To prevent aberrant feedback loops from escalating into epileptic seizures, they cut the corpus callosum, dividing the two hemispheres from each other. The procedures were largely successful, and the grateful patients carried on much more normal lives until some more experimentally-minded neuroscientists asked if they could start doing tests on how they were managing. To everyone's amazement, they found that the sensory and motor systems of each side of the body were functioning at a reflex level completely separately from each other, *without one side of the body knowing what the other was doing* (perhaps a bit like an octopus's eight legs, each of which has its brain, each also connecting together more centrally). Beyond reflexes, however, the experimenters found that by giving contrary instructions to the left and right visual, auditory, and tactile sensory fields they could make left and right hands compete with each other, still without them recognising each other. Had they created two people with two wills?

The vital question was answered when one of the patients said indignantly to the experimenters, "You're trying to make two people out of me!" This was not the result of verbal dominance of one side over the other. It could only have been said because the informational integration of both hemispheres for decisions and personal will was occurring deeper than either hemisphere, in the core limbic brain. In normal living, the corpus callosum connects the sensory fields and fine motor planning across the midline, but the emotive integration of a person is organised at subcortical levels. A person's body posture, balance, physiological organisation, facial expressions, and general coordination occur at this level with relevant feedback from the wider ecology. Choice options are weighed here and balanced as values in a feedback process that recruits the imaginative memorising of each hemisphere via their own cortico-limbic feedback loops. It is in the core limbic brain, at heart level, that the person is mostly and informationally integrated to be an experiential presence and active agent in a shared world. Important orientating, analysing, and categorising inputs are received there from the cerebral hemispheres. However, these *inform* the options for values-based choices at the core limbic brain level. The fulcrum of personal balance as an agent in the world is at heart level.

Through extensive experimentation, the different and complementary roles of the right and left cerebral hemispheres have been mapped. The left hemisphere's networks (left brain) are analytic. They differentiate and diversify one feature of life from another, and attach symbolic language to them as names, mostly as nouns, that can be manipulated by verbs connecting primary subject and object (or subject and predicate). Language increases the specificity with which memories (diverse, overlapping, *dual-aspect conceptual neural networks*) can be reactivated. This increases the powers of comparing and contrasting, so that exploratory activity can be planned. The concepts generated here are boundaried, more digital, with an imaginative orientation outwards towards empirical activity in the world. The main contribution of the left cerebral hemisphere to the core brain's activity is a focus on *change* of the ordering of digital objects.

The right hemisphere (right brain) by contrast tends to be more synthetic. Its networks are silent *without language labels*, although I like to play with the idea imaginatively that they provide the hum, sometimes verbalised, in the background of thought. The right-hemisphere networks tend to gather concepts and notions into categories ordered by their shared qualities. The right brain's world adds the feel for context, times, and spaces to the patterning of life's experiences. It adds presence to the objects named in the left hemisphere, and it adds categorical continuity through change. It informs the more artistic side of human nature. The concepts generated here are more wave-like, with boundaries seeming to be where one concept meets with another, or blends with it conversationally, as if across a fence between neighbouring homes. The right brain mainly contributes *form* to the core brain's balancing of values.

The right brain also tends to initiate withdrawal responses into an inner world. The extent to which that inner world seems separated from other people's and the wider ecology of life may depend on how actively it connects across the corpus callosum with the left brain. In a healthy balance, the two hemispheres would be completely coordinated with each other across the midline. From the left an analytic, quantitative set of outputs adds a *focus* on life, while the right synthetically categorises a socially qualitative *context* for a mindful walk through the shared world.

The inner ecology of a person is thus a complex adaptive system that receives, as its socially related heart, a continuous informational stream of change guidance from the left hemisphere, and likewise of contextual form guidance from the right. These balance in the person's valuing of life's movements, filtering features of the *relatedness* of that person in the wider social and physical ecologies. That incoming reception of the world into the heart of a person appears experientially for the person as their corona of emotive affect, radiant from their anticipations. Thus, the changing relatedness of forms that is triquetrally balancing in a person's inner heart keeps life moving, and steers awareness, thought, will, and emotion.

The triquetral icon of self-organisation in systems

I have used the triquetra firstly, in Part I, to describe iconically how informational principles of organised movement can create an orientation horizon for social interaction in the conscious mind. Pattern recognition in the cerebrum is the active interface of the embryonic ectoderm with wider ecologies.

In Part II I am drawing attention to how the whole person's inner ecology can be represented iconically in the central zone of a triquetra. Three different types of specialised input balance there and integrate to keep life's movements constructive. In Part III we shall see what happens when these three specialised inputs do not integrate completely in the inner ecology.

Below the orientation horizon in the inner integrated ecology, a person's awareness and sensitivity is activated where wider ecological systems overlap their inner movements in their liminal zones. The yellow (pale in greyscale) corner of a triquetra is an iconic ligand inputting and mutually feeding back coherence with other diverse systems. Relationality is real here in its personal movements. Relatedness is immeasurable, however, because in the core limbic brain it has not been converted into an object that can be compared with another.

Nevertheless, relationality is the process ontology of a person's values.

The living body is a self-organising manifestation of this quality of relatedness, bringing together all three ecologies at a wider and deeper heart level than the consciously comparative mind can appreciate. Embodiment is the substance of coherent mutual feedback between three ecologies. Our understanding of this is partial and selectively filtered.

If a person's inner heart is troubled by worries, traumas, disappointments, or hurts, especially if whirlpools of loss emotions are maladaptively simplifying emotional chaos, then the emotional turmoil there will liminally influence wider systems by disturbing the filtering of the feedback loops of life's movements. Through hormone systems in the bloodstream, and autonomic nerve supplies, the physical heart's pumping rate and strength is highly responsive to changing levels of neurological activity in the core limbic brain. Physical sensations feed back from this activity, so there are multiple routes by which informational feedback can affect the progress of intra-action in changing situations. The heartrate would be showing beat-to-beat variation in its rhythmicity to reflect this constantly changing network of informational feedback. If that variability reduces, so that the pulse becomes fixed and regular, then the physical heart has disconnected from its complex entanglement in these feedback loops, which is a sign of illness or heart disease. Therefore, the functioning of the physical heart, its biochemistry and biophysics, is closely connected with that in the core limbic brain, so much so that the notion of inner heart should be imagined to be distributed through every cell of the body in its communications patterning of activity as a whole.

Complementarity of analysis and synthesis in the cerebrum

That which does ascend through the filters into the cerebrum encounters both an analytic and a synthetic process, which need to seek their own balance. These two qualities of informational processing to put order into data are differentiated into the left and right cerebral hemispheres respectively. Each contributes its patterning through the left and right sensory association areas forward to the motor-planning areas, and down to the core limbic brain. Analysis and synthesis both contribute to the perceived quality of life above the orientation horizon in conscious space, time, and experiential presence. The left hemisphere is analytic; the right hemisphere is synthetic. The predominance of left or right hemisphere contribution to the person's unique personality and character is balanced mainly in the core brain, where it collides with the social feedback loop represented by the yellow corner.

The two hemispheres are mutually connected at every point across the midline by fibres running through the corpus callosum. Balance between analysis and synthesis in a person's relationality could be perfect if there is complete coherence there. People tend to be born with a personality predominance of either one, but neuroplasticity allows educational development.

The liberating effect of learning Emotional Logic works at this level of cerebral lateralisation and specialisation, as explained in Chapter 1, in the section on unity being balanced in the core brain, and diversity generated in the cerebral cortex. Learned behavioural habits of managing grief emotions are built wordlessly into the right brain's synthetic and categorising roles, assigning qualities and context to experiences, which affects subsequent anticipations. The left brain is more involved when first recognising certain behaviours, assigning names to them with relational frames and boundaries to the observations so that responses can be managed in a more focused way. If, for example, someone has had a bad experience of anger in their past, or has a natural tendency to feel guilty when something needs to change, then the left brain will recognise and name observed behaviours as anger or guilt, and the right brain will perhaps categorise 'it' as 'dangerous', 'things to hide from', 'something wrong with me', or 'reasons to play dead or pretend I am not here and to disengage'. Using the kinaesthetic mapping tools (card patterns, loss reaction worksheets) engages the right brain more effectively than talking about emotions. Comparing patterns with the theoretical Life Cycle diagram and other Emotional Logic materials provides the rational and word-based input that activates the left brain. Reading the useful purposes of unpleasant emotions as integral parts of a healthy adjustment process, which gives meaning to seemingly random experiences, integrates left and right hemispheres in a way that can recategorise the observed behaviour and thus alter anticipations. This is how the liberating effect becomes embodied to renew and refresh personal engagement with life.

Releasing informenergy from whirlpools of loss emotions

This liberating process is especially powerful for releasing informenergy for adjustment and growth that has been trapped in whirlpools of loss emotions. In Emotional Chaos Theory terms, renewed understanding leads to a bifurcation of a maladaptive attractor state. This energises mutual synergic connection between the inner, physical, and social ecologies, resulting in a Butterfly Effect of fractal scaling that repatterns both the inner ecology of personal identity simultaneously with the social ecology of qualities of relationships, and the physical ecologies of health and environmental order.

Thus, an informational view of emotion in partnership with awareness, thought, and will can energise life-enhancing responses. When renewing anticipation patterns, people feel empowered to try *rationalised* risk-taking and experimentation, not careless risk-taking. They can more effectively search for responses that will break the old unhelpful emotional habits of feedback. As a consequence, people can build more responsive and enduring relationships, forming stable domains of groups that include diversity in order that each can enrich the other's fullness of identity.

Renewing a mental framework therefore can add new meaning to the feelings of emotion. This mental framework can be taught to others in a social ecology, making emotional messaging more comprehensible and reasonable. Other people may then seem less threatening when people are equipped with depth relational skills, so time can be given more willingly to discover each other's values. The range of options for agreed responses increases. Less traumatisation reduces unhelpful memories. Joy may be quietly born within. Empathy with the grief that other people's values have brought them can shift conversations to a new level, where previous enemies can discover their shared humanity. People who differ may cautiously become friends with their 80–90% shared values.

Chapter 6

Energy and Informational Processes Are 'Informenergy'

Personal energy is physiological order

The archer and the timing of choices

The strength of an athlete is not due to muscle bulk alone. The athlete's training coordinates inner organisation. Every relevant muscle fibre contracts in a synchronous pattern when the central command arrives from the nervous system. The power of choice (mental) to release energy (physical), and the power of choice to resist releasing it, both equally change the world, the lives of the spectators, and the written history of sport. In the example of the athlete, the relational quality inherent to all movement at that time is competitive rather than cooperative, although team cooperation and morale does help individuals to perform maximally.

There is a Zen paradox of the archer, who is also the bow and the arrow. All are embedded as one integrated unit in the wider life-context. This is a helpful image to learn about complex adaptive heart-level systems. It may need to be revisited over years, as one's depth of understanding about the informenergy of choice and of timing progressively makes more sense. The terms 'personal energy' and 'informenergy' will be increasingly used as this book progresses. Personal energy is an organisational state of physiology. The archer has a highly organised and targeted choice to make that integrates all three ecologies: physical tensioning, relational purpose, and inner balance. Choices are made with focus and timing *within relationally systemic* inner and social ecologies. Mental distraction reduces personal energy; emotive preparation in the longer run-up or training timeframes required to prepare for performance enhances personal energy. Focus and context are both more relational than people commonly think.

The term 'personal energy' is carefully chosen. Its intention is to find a middle path avoiding two extremes to which people's comprehension of 'energy' may drift. One extreme is to understand the energy of life as purely a physical, electrochemical phenomenon. The other extreme is to understand the experience of energy as an esoteric, pre-material, vital force. In the former view, random disorder is at the root of life, which needs self-will to control and organise. In the latter view, if people are to be healthy or effective, even in sport, they need to choose to harmonise their lives with a pre-existent level of order that they fall short of. The double triquetra offers a third creative way to picture the personal energies of movement and stillness.

The double triquetra is an icon of the balance and synergic conversational connection between diverse states of life. Balance and synergy are achieved by focus on the *relatedness* at heart level of one substate (one triquetral change or movement) embedded within or connecting responsively with wider dynamic systems (the other triquetral ecology of transforming movement). Responsive connection creates a liminal zone of complex overlap that *is* relatedness. Patterns of changing relatedness emerge as internal stability and external physical order. The athlete springs into competitive action. The archer chooses to release the arrow. Simultaneously, the inner and social ecologies reshape around the choice of timing.

In the triquetral approach to cosmology, spirituality is one and the same as qualities of relatedness within the Bigger Picture of life, being moved or responsive beyond the locally known environment. By calling triquetral self-organisation and timing *personal energy,* I am creating a liminal zone in the relationship between human beings and all other biosphere living forms who, or that, also have synergically stable inner organisation of their physiology. I may be at risk of extending the notion of personhood too broadly in readers' minds if the concept of personal energy is attributed to all life-forms. That is not my intention. When trees and rivers are also considered persons, the notion of spirits and spiritism can distort the qualities of relatedness in the Bigger Picture. These can lead, I believe, to maladaptive ways to manage emotional tensions. Worship of idols, and the appeasement of fearful mystical power hierarchies, are not the way that New Science leads spiritual development; neither is the rationalising Enlightenment Science, its opposite, a recommended direction for personal development, because it reduces the wider spirituality of life to mere dust and randomness. However, triquetral resonance of movement between diverse self-organising people is a concept that liberates individuals from these misguided searches for security. True personhood is about the *choices* that people make in the complex liminal zone of their qualities of relatedness when balancing options around their values. We need to assess how much power of choice various other self-organising features of our environments have, if we are to decide that they are *personal* substates. Awareness and respect for life in all its varieties does not require that all life, or all environmental features, be considered conscious if they are not comparatively choice-making. They definitely are responsive and possibly aware parts of life's rich diversity, and should call forth kindness and mercy in our responsiveness to them. However, much of the power of choice in this small corner of the cosmos rests with us as human beings, who *are* comparatively conscious and responsible as we make decisions.

Human beings always have a power of choice, if not about what they do, then always about how they do it. It is a foundational assumption of a triquetral view of personal development that every human being has personal value, and dignity as such. Cultures vary enormously in how they diverge from this balanced view, but the conditions for conversational exchange by choice between cultural, group, and personal identities are being made explicit here. Then life-enhancing choices can increasingly be made during times of environmental deterioration.

Relatedness and synergy scale into energy

From here on in this book, the notion of *energy* needs to bring to mind the double triquetra with its relatedness ligand informationally joining two self-organising movements. The relatedness ligand shows the resonant synergy, from which energy emerges by further entanglements of synergic vibrations.

When a triquetra in a double triquetra is being used to represent a conversational person, however, its central zone may be imagined to be a moving emotive pattern of affect. That moving pattern can be mapped with loss cards (Figure 12) and tick patterns on loss reaction worksheets (Figures 14–16). There will be extra joy-related emotions also, making a pattern that is the preverbal corona of inner ecology movement shaping responsiveness within conversations.

It is the list of hidden losses in the left-hand column of the loss reaction worksheet that connects heart and mind into a rationalising partnership to manage change and make life-enhancing choices. The examples shown here have seven named losses, previously hidden within the shameful experience of making a mess of an interview. When Emotional Logic coaches converse with people who have been traumatised by serious abuse histories, three A4 pages of named losses may come pouring out, with multiple whirlpools of loss emotions appearing in the tick patterns. People sometimes hold their loss reaction worksheets in both hands and say something like, "Well, no wonder I feel bad!" Our coach might reply, "If you didn't feel bad, there would be something wrong! This is the grieving you. You only grieve if you have loved, so that makes you more of a human being, not less. Join the human race! The more you can make friends with your unique way of grieving, the more choices you will discover to develop a settled and stable personal identity with which to rebuild relationships."

The conversation is an exploration of how each person discerns and responds to the other's inner patterning in their shared environment. The dynamic between them, in their liminal zone of relatedness, will inevitably fractally scale from their interacting *patterns of emotion* to release personal energy from their internal stability or instability into the shared world. That release spreads harmony or conflict. The key concept to take forward from here, more foundational even than energy, is conversational synergy, and the fractal scaling that moves energy from it into wider ecologies.

Synergy is the resonance of feedback. It is the potential that can convert into kinetics. Personal synergy in some healthy relationships brings mutual changes of relatedness that generate a sense of *presence together* on a higher level of order. That resonance is the fusion of two potentially isolated individuals into their *personal* identities, as friends in relationship. In a prime relationship this can develop further into a partner identity, which some people will take to a pattern of commitment called marriage. Resonance can also develop in groups of individuals, as people discover a synergic fusion as a group identity. As group identities emerge, the group values also influence the kinetic action that its individuals could not imagine they might do on their own. Mutual understanding in any of these synergic states can lift heart-level communication to a new level of order, sometimes beyond the orientating framework of space and time and experiential presence in the locality of a meeting place, as in some worshipful or ecstatic states of being.

Fusion and fission as energetic movement

Relational fusion is a process whereby two or more stable states synergically coordinate into a more integrated pattern as a new dynamic form. Energy is released from that process into the wider ecologies. Fusion reactions in the sun release energy for change that warms the planetary earth, and enables life-forms to emerge from dust and water. These life-forms further fuse their patterns of relatedness in a social ecology that may release creative energy into the wider ecologies of life, seeding more growth or organisation. People can become sources of light, warmth, and coherent stability for each other. The fact that social systems are open systems, some of them shaped by personal values, reflects the fact that the cosmos is an open system shaped by the same core principles of organisation at every level.

The opposite process is relational fission. A single integrated state, taking the nucleus of a large but unstable atom as an example, may split into two self-isolating parts, similar to a marriage separation or a group split after disagreements. Fission also releases energy during repatterning of the form, which had been temporarily held within parts of an unstable system. Technology has been able to harness released fission energy in tightly regulated nuclear reactions in power stations and warheads. It is the practical outworking of control attitudes towards nature. However, this requires a parallel condition of the human heart. The resource to generate that energy relies on having chosen to create instability in nature. Research has generated new unstable isotopes specifically to break them down again when chosen, and to learn how to regulate their exposure to each other so that the degree of instability and release of energy can be controlled. This Enlightenment Science approach to technology is based in a belief that the human inner heart and the mind that is moved by it are inherently stable enough to regulate it.

The principles of organisation that can contain accumulating nuclear instability in nature can also, to some extent, contain the accumulation of unstable emotional energy in entangled physiological states at a human being's heart level. Under changing environmental and relational conditions, however, these personal states may release their entangled energy in unregulated ways from the complex patterns that can be mapped in the inner human heart, especially from Anger-Depression whirlpools. That entangled energy may fractally scale as an explosive outburst in ways that disrupt life, sometimes disastrously. Such an explosion of misunderstood emotional energy may simultaneously disrupt the individual's mental health, and disrupt their social behaviour. Sadly, unstable emotional patterns in the human inner heart that trap personal energy to adjust in life have an opposite effect to morale. Whirlpools of loss emotions other than the Anger-Depression one can collapse individuals into behavioural and mindful states of isolation and demoralisation that make them vulnerable to relational manipulation by others, or drive an urge to control or hurt others.

The personal development strategy being presented here places the same *informational movement* both at the heart of human social reasoning, and at the core of natural environments and all emergent life. In human social reasoning, informational movement emerges as emotional states that need interpreting. But for health and hope to grow, they must not be interpreted just in the context of the individual's unique past experience. To make useful sense of them, a Bigger Picture of relevant change processes is needed that creates a liminal zone of relatedness among all three ecologies. The double triquetra shows how an individual's experience may then stabilise into fusion energy states in a Bigger Picture of relatedness.

The resonance of mutual understanding that could be conversationally aimed for can scale fractally. Its synergy can reintroduce stability into an open system of life as it helpfully releases fusion energy for growth.

Triquetral people and informational links with the land

Our small, blue and green, but increasingly dusty planet needs looking after over the next 30 years. The problem is getting people to cooperate to survive with sensitivity and adaptability when local immediate needs and rivalries dominate people's thinking. Some still want to thrive at other people's expense. Much is already happening at the financial, political, and business levels to prepare incentives that encourage innovative adjustments.[1] However, the losses of lifestyle and comfort for individuals over the coming decades are going to be enormous, for example when sea levels rise enough to require population shifts, and the economic system based on insurance pay-outs collapses. Values need to be explicit.

The concept of becoming 'a people' cannot be separated from the land. But the land seems a remote concept to city-lovers, for whom the land's economic produce comes pre-packaged in supermarkets and mini-markets. Individuals in city mentality can, however, develop firstly as *persons of a people*. They may grow then to think naturally of a wider liminal-zone picture of relatedness and cooperation around explicit values when facing change. If not, then conflict, entrenchment, and isolation will tip life into turbulence, escalating into chaos, and then entropic decline, probably associated somewhere with a bland imposition of totalitarian rules.

Triquetral principles of organisation are simple to intuitively understand and balance when conversationally shaping the future in agreements. When people interpret loss emotions not as signs of weakness, but as vital information about personal values, and as potential energy for adaptability to become a people, then emotions can inform anticipations for stability in order to navigate change. Anticipations develop sensitivities that can focus on life's emotive synergic potential for renewal. That awareness can reactivate repeatedly over the next few decades in whatever ecologies emerge in the global process of change that we currently face.

A flight over the landscape of relevant philosophy

The process of mental separation from the land has been ongoing for millennia. The New Science is restoring the potential to rediscover that link between humanity and the land in a cosmos that innately births life wherever relational stability makes it sustainable. We see life discovered in the most unlikely environments of this planet earth. Statistically, it is highly likely that there will be life-forms discovered in other localities of the universe if humanity does not extinguish itself on this earth first. Understanding the New Science of emergent order in adaptive complex systems can help people to stabilise life on earth now in new ways that can continue to liberate the spiritual potential of the human being's inner ecology while also respecting the established order of the physical ecology we share. This evolution includes bringing societies out of a long history of Stoic and Confucian exclusion of emotion from the attribution of honour. Restoring respect for intelligent kindness and enduring love is vitally urgent.[2]

To recognise the analytical perspectives described in Part III that veil this conversational reality revealed by the New Science, it will help first to grasp hold of some of the notions that we are going to shake.

Dualism

The Enlightenment 'Age of Reason' may be considered to have started in 1662 with King Charles II granting his Royal Charter to the Royal Society of London "whose studies are to be applied to further promoting *by the authority of experiments* the sciences of natural things and of useful arts" (italics mine). Sir Isaac Newton was one of the earliest fellows of the Royal Society of London, elected in 1672, and its president from 1703 to 1727. The *authority of experiments* has indeed immeasurably extended human understanding of the natural world. The 'useful arts' of technology have indeed raised living standards for many.[3] However, modernist science leaves unstated its core assumption, that conversational *agreement between people* is needed to *limit* the ecological conditions that define an experiment.

As a consequence, the conversational reality of scientific exploration has been lost to sight behind a veil of the dispassionate study of, and publication about, material objects. Firstly, this attitude and conversational agreement was applied to natural physics, for example to gravity, and gases, and coloured dyes. It was then more boldly applied to the human material body, the brain, and genetic chemistry. Then the notion of controlled experimentation was extended physically to the behaviour of animals and humans in *experimental psychology*. The dualism of Descartes (1596–1650) supported this notion that a human being could become a material 'it' under observation. This became the dominant authoritative attitude also of other rationalist philosophers during the Age of Enlightenment, who saw the clockwork universe as a closed system to be studied by an observing rational mind. This period of intelligent, rationalist philosophy came widely to the same conclusion, that the dualist mind could have no direct knowledge of the matter under observation.

Personal identity was thus ontologically separated away from physical matter. This has been briefly mentioned in Chapter 3, 'Becoming Informationally Orientated'. A reminder now about the difference between object-focused philosophy and process-focused philosophy will help to lay a firm foundation on which to stand, like an observation platform raised up to the level of the orientation horizon. This platform will be needed when we move on in Part III to see how neocortical analysis can *veil* the heart-level awareness of balanced reality. The various veils allow unbalanced values to hinder transformation, which is probably not a life-enhancing process when the global population faces unrelenting transformation of the land under its feet.

Concepts have moved on greatly from the 'modernist' and rationalist Enlightenment agenda of the seventeenth to the twentieth centuries. Newtonian modernist science is based in a notion of the momentum of solid objects on the move and colliding with each other. The paradigm shift currently ongoing in the twenty-first century replaces the whole notion of solidity with energy repatterning. The paradigm of this shift is the widely known notion of wave-particle indeterminism. The fascinating mystery remains, however, how the indeterminism of waves superimposing on each other at that quantum level of the cosmos can transition into the particle states (and vice versa in particle decay), and how these events spread their impact *deterministically*. Cause-effect sequences among particles *do* scale up predictably. Matter on its own *does* do what we choose to push it to do, provided we observe it in a linear time direction from past to future with measurable momentum in a closed observational system. Wave states, however, reduce that determinism as particles decay, for example in friction.

People who have been educated in schools and colleges into a modernist, dualist view of science, and who believe in the predictability of matter, will need to transition through an emotionally turbulent time of uncertainty if they are to grasp the full joy of liberation that can come with valuing the New Science of adaptive open systems and living renewal processes. It starts with ceasing to quote inappropriately that the Second Law of Thermodynamics predicts the decay of order into entropy, and that the First Law of Thermodynamics declares that energy is conserved and cannot be created. This is because both of these Laws apply *only* to closed systems. The systemic, adaptive life we share is an open system in an expanding, open universe. We need to open our minds accordingly.

The emotional turbulence might end when realising that the truth criterion for 'science' has not changed during the transition to open-systems thinking. *Replicability* is the truth criterion of science. And replicability establishes *predictability,* which empowers human beings to inform their choices of action in ways that do not rely *entirely* on intuitive divination. The New Science introduces choice about how much to modify the intuitive discernment of heart-level trends. It does not devalue intuition, but adds to it a consciously comparable level of awareness with computer-assisted algorithms of the impact of multiple known and measurable factors. The problem will surface, however, when human beings make themselves subservient to information (noun) being processed in the physical ecologies through technology. The need for a choice between options and a decision for action remains. That power of choice remains at human heart level, in a complex patterning of informenergy that connects the human being into the wider movements of the cosmos in a way that cannot happen with robots.

Not all factors that influence human heart-level decisions are measurable. Computer algorithms capture only some of the variables of social life emergent from the land. The advantage of algorithms over the Enlightenment Science dualist philosophy of solid particles with momentum, however, is that the mathematical formulae of algorithms are the language of *process,* not of form. The problem with algorithms starts when people believe that the product of an equation is a truth about life. It is not. It is a product of conversational life at every level of relatedness within one cosmos.

Human beings are intra-active within the processes of life. Life processes are informationally rich networks. We share these liminally with others. The truth criterion of replicability is achieved not by controlling the physical conditions of the experiment, but by mathematically modelling the multiple systems of dynamic reality. The partial contribution of New Science to living truth emerges, however, as *measurable probabilities* that restore the goal of *contributing* to predictability. This development of the truth criterion balances determinism with potential variability. There is great benefit in this, but also an associated great danger arising from the mental tricks that can hide the conversational, synergic depths of reality. People can deny their heart-level reality, and become robotically subservient to algorithms. Slavishly following protocols dehumanises life. Educational training is needed so that people value intuitively managing the 15% of situations that do not fit the protocols. The New Science opens a risk that information (as a noun, as if 'in' a protocol) might replace form (as a noun, as if solid matter) and spirit (as a noun, as if esoterically pre-material) as the dualist idol of the future that is slavishly worshipped.

Knowing the triquetral balance in a conversational reality is needed to resist this fall from creative relatedness.

Monism and dual-aspect monism

Philosophically, the main alternative to dualism has for a long time been monism. Rather than acknowledging two essences (from the Latin *esse,* 'to be') of mind and solid matter that cannot know each other, monism is the notion that there is only one 'essence', and that all the phenomena of life are diversified features of this essence. A range of ideas has emerged throughout history about the possible nature of a single foundational substance or essence to life, called 'substance monism'. These views are common to many Eastern and Western spiritual or religious traditions, but they share a common weakness. The human being is seen as a rather passive participator in living processes where diversity is assumed *already to exist*. To enquire how diversity arises from the monistic essence is left as a mystery. To sustain monism, diversity simply is.

Neither dualism nor substance monism can provide a robust account for the diversifying and reintegrating relationship that we know experientially between our socialising body and its imaginative mind that can drift off into daydreaming. Neither can they account for the mutual adaptability of body and mind when environments suddenly change, nor can they account for the human capacity to play with the breeding genetics of plants and animals to transform their features for human economic benefit. Pushed to an extreme, the proponents of monistic philosophies tend to deny the existence of either matter or mind, claiming that one or the other is just imagination. Following Spinoza (1632–77), and more so in recent years, there is an ongoing attempt to create a philosophical middle ground in monism, which describes mind and matter as different aspects or modes of a single substance of life. The exploration among philosophers of this 'dual-aspect monism' has been widespread, but it still has divided into the 'positivist' camp, who claim that there are ontologically real, self-existent things of some sort, and the vitalist camp who claim that a monistic essence lies *behind* the appearances of all things physical and mental. During this last century, this has led to a more radical break from the positivist focus on forms, called process philosophy, which focuses on change.

Process philosophy

Process philosophy is not new. It started on record with Heraclitus of Ephesus (approx. 535–475 BCE), who believed change and flux was behind all that seems constant in the universe ("Everything flows"). Its more recent revival is attributed mainly to Alfred North Whitehead (1861–1947), whose thought challenged the ongoing rapid development of subatomic particle physics.[4] He offered physicists the ability to think more in terms of the emergence and dissolution of forms as a process, rather than as a splitting or decay of self-existent ontological substance-energy.

Every physicist has become a philosopher as a consequence, having opinions and models to fit experimental findings into a secure order. Positivist views of ontology had been based on the First and Second Laws of Thermodynamics, viewing the universe as a closed energetic system with no new creation or renewal of energy. Process philosophy offers a different view of reality, as a permanent state of 'becoming'. This is consistent with the strange findings of quantum-level physics that do not fit within our orientating concepts of space, time, and substance. While process philosophies tend to focus on energy and waves, they still do not give a good foundational account of the *diversifying relationship* between *mental and material processes*. There is a trend to overcome this by talking about 'consciousness' as the essence of life's becoming. However, this use of the term may be only a euphemism for responsiveness. Equating universal consciousness with universal responsiveness does not explain how matter is deterministic in its cause-effect behaviour, while mind can go on flights of fantasy. Another refinement of process philosophy has been needed, which may be offered by the triquetral process cosmology being described here.

Process triunity in open systems

That refinement may be to expand the notion of monism into a *living monism*. Monistic *unity of movement* is knowable as *triquetral principles of organisation*. A monistic dynamic in movement is a *triune process*. Diversification is inherent to triquetral process unity, because networks produce localities among informenergy movements. Diversity becoming, and diversity dispersing, are equal features of triquetral movement networks. This process is foundational for living forms to scale up and disperse over timeframes among energy-matter transformations that may be known as the land.

Process triunity extends from quantum to universal as a uniting informational cosmology. It is iconically represented in the double triquetra. It is a conversational cosmology of resonance and synergy at every level of physical, social, and inner ecologies. The human mind diversifies its ideas when the processing in neural networks reactivates partial features of stative data in memory networks. Bits of memory become *incorporate* as informational feedback ideas. The light of consciousness is sensitivities remembered. Flights of fantasy happen when the same informational reactivation occurs with non-body image processes.

The same principles of feedback organisation of movement may apply below the orientation horizon, at the quantum process level at which material objects are also synergic changes of quantum relatedness between localities within a network. Stative potential becomes converted or incorporated into objective patterns. The result above the orientation horizon is a mental impression of embodiment with boundaries. Below that horizon there may be an intuitive awareness of coherence in something bigger, but with less memory association.

An inner-outer personal reality of embodiment can thus be recognised at the orientation horizon. Patterns of emotive progression (affect) arise from heart-level mismatches between feedback cycles. The inner human heart can become troubled or at peace in the world, not only by coherences in the social ecology feedback loop, but equally by neocortical feed-down loops activated by pure fantasy. They both use the same informational process triunity to initiate movement. However, one is a movement into the material world, and the other is a movement of ideas back up into the analytic but now dissociated cerebral cortex. These feedback loops may, of course, complicate each other. Fantasies can lead to physical activity, and vice versa.

Triquetral emergence of order does not require any core essences or primary ontological substance. Its source, its core or seed, is *mutually interdependent co-origination in feedback*. No one feature of triunity can have any reality in or of itself without a coherent dynamic that involves the other two.[5] The core of life *is* mutual coherence; and mutual coherence *is* substantial and understandable. From this core, unending dynamic, a triquetral conversational process mindfully progresses with slight variations in iterative feedback. These slight variations in coherent movement can fractally scale and replicate into numerous systemic levels. Thus, diversification can emerge from quantum indeterminacy into the *triple feedback ecologies* of life, in the mindful phenomena of material, social, and inner feedback cycles.

Triune means alive or living

A triquetral cosmos is *'a universe with emergent and diversifying life'*. Cosmos is movement. *The* cosmos as an ontological noun is an orientated myth resulting from activity in the temporal lobe sensory association area. Triune means *living*. A triune person is alive, adaptable, and relationally responsive with others who differ but who are able to remain in unity of process.

As mentioned before, however, the mutual coherence of life *is* substantial as an enduring process. It can scale up into atomic technologies and large-scale human engineering projects such as bridge-building that do have predictable and deterministic processes. These processes maintain their form within a known range of flexibilities, but they need maintenance, and will eventually wear out through friction. When five compartments of the *Titanic* were breached by the iceberg, it was certain that the ship would sink despite all the prayers.

The etymological source of the word *substance* is important here. It comes from the Old Latin and Old French for *under* and *stand*. It refers to that which is firm enough to stand upon. In the English language, the word *understand* is used exclusively for the mental phenomenon of coherence, when a thought in mind connects with memorised ideas or events. The separation of physical substance from mental understanding is an example of embedded dualistic thinking in our use of everyday language. Process triunity can offer another concept, one that allows useful diversification into material and mental, but which may help to overcome the way that ontological dualisms prevent humanity from becoming truly alive. It is the concept of informenergy.

Informenergy moves people in open systems

Just as 'a substance' and 'an understanding' are objectified dual aspects of a common underlying triquetral process, so too are energy and information objectified dual aspects of that same triquetral process. The systemic process is pre-orientation in space, time, and experiential presence. Our experience of life's movements emerges out of it. Consciousness above our orientation horizon constrains the open-system connectedness and potential awareness of our inner hearts.

As mentioned in Chapter 3, specialised languages develop in local contexts. These languages develop a vocabulary that brings the core principles of movement into practical, empirical usage among those people who share an interest in how to set that area of life moving. The vocabulary diversifies to assist analysis of how a local system of interest works, how it breaks down, and how it can be fixed. It is a vitally informative process, provided that the findings within that specialised field of knowledge do not become a closed system, inaccessible to the wider community. Practical application for the benefit of the wider whole completes the specialising formation of diverse personal identity, group identitics, and cultural identities.

Translating specialised language into generally understandable concepts is called 'popularisation'. Specialised terms, such as those summarised in Tables 3.1 and 3.2, have gathered their own relational frames for those who use them, which may not easily cohere or resonate with those meanings for a general readership. The tragedy we face as a global population at present is that practical application of dualist science and philosophy has embedded relational frames for words that people are unaware of. The practical outcomes for people who are interested in people tend not to be those that liberate responsiveness and adaptability creatively, but cynically to improve sales, or to attract cultish 'followings' that are relationally closed.

Informenergy is a core concept of adaptable relatedness. To some extent, breaking out of dualist habits of thinking that are currently unrecognised will inevitably generate grief emotions for the loss of certainty and predictability of the familiar. But grieving is a motivated mode of love. In this case of letting go of old habits of thinking, it is their love of 'getting it right' that people may feel is challenged, believing that 'getting it right' means relatedness, when these two concepts are, in living process fact, different. In a triquetral cosmology, 'getting it right' means adaptability in relatedness. On the other hand, people who are already grieving because they know that something is missing in their prevailing society, but cannot define what that is, may have an outburst of joy on hearing that there is another way to see things and to connect with others. Enlightenment Science, and the old closed cultish belief systems associated with it, have left many people feeling isolated and lost at heart. By becoming familiar with how grief leads to a growth cycle, an encounter with process triunity can become life-enhancing.[6]

Informenergy is embodied life in open systems. The concept of informenergy starts with a large-scale view of an open system in which there are many smaller dynamic systems, often called elements. Living forms emerging among these elemental states share in the accelerating expansion of the universe. This is a paradigm shift of thinking on a scale equivalent to the Copernican Revolution. The thermodynamic view that a system could cool if it is closed does not apply to the cosmological context for the growth of life. Our ecologies all remain open. There is energy exchange in and out of the physical, social, and inner ecologies of a human being, which is the informenergy feature of adaptability. We shall go on to see in Part IV how the physical ecology of the universe can be an open system, which will be an astonishing idea to people educated into a belief in the conservation of energy-matter. The systemic fact that habits need their fractal range of variability to remain features of open systems applies as much to the habits of structured matter as it does to the habits of group identities and cultures. Everything moves. Everything in the cosmos rotates or vibrates or translates.

Diversifying change becomes self-organising and self-replicating within cooling areas of that universe that include stative patterns of activity. One potential diversified change would be when someone who is self-organising their environment chooses to section off a part of the whole into a closed system, and measures what happens to the nouns they describe within it. But most of the contextual informenergy system of the whole remains open. It compensates for the closed-off sections until they open up again.

This process happens in brains as much as in other physical energy-matter systems. Sections of neural networks close down for a while during rest. The remaining living tissue of the whole body compensates in a range of stative-active ways. While the brain sleeps, the immune system switches into 'healing' mode from 'defence' mode. The memorising systems go into dream mode from 'focus attention' mode. The body goes into 'growth' mode from 'action' mode. But the living system does not switch off. It remains open, so its neuroplasticity can generate new degrees of freedom to explore while other areas become habitually settled into memories and physical development. That's life.

Informenergy is holomovement in implicate order

In the 1980s David Bohm wanted to prevent the developing quantum sciences from replicating the atomist tendency to look for smaller particles or packages as the basis of reality.[7] Approaching from the direction of mathematics and physics, he proposed the notion of an implicate order at quantum or pre-quantum level that is in continuous movement throughout the whole at a level prior to the emergence of an explicate order to reality in space and time. He believed this implicate order was not chaotic, but could be described algebraically.

The triquetral cosmology being developed here is a way to iconically picture the holomovement of that implicate order. No mention is made of algebra, but the hypothesised triquetral icons could be expressed in algebraic terms that would allow experimental checking of their predictive value. The approach being taken, however, is from a neuropsychological starting place. It considers how the individual's brain can informationally construct an explicate order in a consciousness that primarily has a social function. Consciousness emerges out of sensitive awareness to orientate conversations or interaction with the land for survival. The brain is not built just for analysis or individualistic pleasure, but the explicate order it creates can become very self-absorbing.

At a human inner-heart level, however, I am proposing that we human beings, and indeed all living creatures, are more naturally at home in the implicate order. And more, that movement in this implicate order emerges into human experience at a heart level of emotive affect, which secondarily drives our thoughts and inclines our minds to focus on areas of individual interest.

The purpose of this book is to open people's hearts to rethink their anticipations and priorities as our shared environments deteriorate in the coming decades. An association of the human experience of informenergy with David Bohm's implicate order will be considered in Part IV, but its significance is for the growth of a renewed humanity, not merely for technological advance in the field of information processing, which may dehumanise life. The term 'informenergy' is potentially more life-transforming for a growth in humanity. It can help to open the closed systems of individualism to restore systemic thinking, healing, wholeness, and fullness of life.

Closed-system communications are robotic

In Chapter 3 I explained how scientific Information Theory has arisen and mushroomed beyond Shannon's original purpose. It is an example of how his purposeful but dualist, closed-system thinking can go unrecognised in the midst of the excitement generated by the rapid growth of new technologies, and the ease with which cultish 'followings' can be attracted by use of unboundaried marketing psychology in social media. Of course, much open-system benefit is also gained through use of social media by people of sound mind and stable heart. Life-enhancing choices are made at heart level, whichever way they are communicated.

Shannon quantified information (noun) in order to measure decay of coherent processes in copper wires between transmission and reception. His measurements, and the mathematics of information, apply to that type of closed system. Scientific Information Theory is not in itself concerned with how the patterning of a transmission is generated, or how that pattern is recognised and acted upon in the receiving embodiment. With the rapid development of aerial transmission and carrier-wave connection to a receiver, the mathematics has mushroomed, but it has remained closed system. This is because the mathematics has necessarily looked more closely at pattern generation and pattern recognition systems in order to define the 'information' (dynamic pattern as *now a static noun*) that is measured, transmitted, received, and compared. The 'information' (noun) is measured by its electro-mechanical effect on a robotic closed system, such as a mobile phone, or a missile in flight. Here we can now see the weakness, or the reduction of life to robotic states, within the Information Technology explosion. The mathematics of pattern generation and pattern recognition compares order (the amount of patterning) with entropy (the absence of any patterning). But entropy is a mythical concept that can exist only within a closed system. Information Technology assumes that pattern generation and pattern recognition systems are both closed systems. *The communication along the carrier wave is dualistically different from those closed systems of transmitter and receiver.*

But—and this is a big but—if we go back to the 1940s when Shannon was developing the mathematics, and before him Harry Nyquist and Ralph Hartley in the 1920s, the technology was developed to enable conversations between people when they are remotely separated by physical distance from each other. The conversation itself is a living interpersonal dynamic of coherent resonance. It establishes substantial change of understanding between two open and mutually responsive personal subsystems. The conversation is bidirectional, unlike a transmission.

The hidden duality of the information revolution for the general population is this. People may imagine or understand their brains to be generators of information (noun), which they 'put out there', and have no responsibility for the impact that it has on relational change. Conversational reality has then broken down into individualist isolation. An isolated person is a closed system, and so can tend towards entropic decay, or to robotic manipulation by misinformation and disinformation. Robots can be switched off into static states. Human beings cannot. When stressed, they tend towards tension, distress, and confusion states of stative isolation. This is not a good condition in which to make life-enhancing choices.

Open-system communications are human

A human being generates words as meaningful resonances with others, although this can be complicated when words mean different things to different people. In one sense, the inner human heart is an open-system word-generator. It is from the inner heart that we tell the stories of our lives, modified by our conscious memorising. Those stories can be told in words, or movements, or artistic and musical harmonies and disharmonies that resonate with the lived experiences of others.

The brain and the immune system working together are an open pattern-recognition system for personal identity. They communicate chemically with each other. People need to explore the bigger context of life that they are patterning internally if we are to thrive. However, the inner ecology may hold us back. Rest is a stative state of healing in an open system of life. During physical and mental rest, repatterning of order renews personal informenergy. Also, taking a brief mental step back from activities occasionally, introducing a short *gap in time* to reflect, can reset personal energy and values. The reflective triquetral state renews anticipations, and restores synergy with wider ecologies for a healthy search. This is a hopeful and stabilising condition in which to make life-enhancing choices.

Chapter 7

A Place for the Human Heart in Physical Science

Part II is examining how heart-level emotion is personal energy in motion. This *personal informenergy of communications* can get stuck, especially when words mean different things to different people. Misunderstanding can lead to broken relationships and emotional loss reactions. In this chapter I aim to prevent polarised misunderstanding about the nature of personal informenergy, by recognising:

- how the history of science has specialist language framed in dualist ideas;
- how the history of dualist philosophies and religions has affected the meanings of commonly used terms about personhood, and;
- how process triunity in open systems allows new beginnings that are consistent with scientific *and* philosophical or religious frames of mind.

Informenergy as *movement from within* casts a new light on old relational frames for commonly used words about matter, mind, and personhood. These can all be relearnt without sacrificing the security of connections previously made. An informenergy view of life simply broadens and deepens responsiveness and adaptability when environments start changing unpredictably, as currently is happening on this planet earth.

Letting go of scientific dualist language

A concept of informenergy is not another object or thing. Informenergy is *co-emerging synergy in movement*. The linguistic problem in English is the necessary grammatical use of the present-tense verb 'is' to describe what a thing is, rather than what it does or how it processes. Synergy requires two states in mutual movement—two triquetral changes that harmonise or resonate. There is nothing static about synergy. 'It' is movement. Synergy is vibrational.

Likewise, the pronoun 'it' is also a potential problem in English. It (the pronoun) is used only above the orientation horizon linguistically, where it refers to a concept that has acquired boundaries. It objectifies changing relatedness very easily. This anticipation of an objective resistance to change can be embedded in the core brain as a personal value, namely that some things should endure even though the perceived object is mentally generated from changing relatedness data, and even that data has been pre-filtered in relation to that value.

The objectifying nature of orientation in space, time, and experiential presence is cross-cultural. In German, for example, nouns are given initial capital letters because they are substantial, and the verb is presented at the end of a sentence to finally dynamise the subject and objective elements of the sentence. The idea that a force (verb) is applied externally to pre-existent substantial objects to move the sentence, spoken or written, is so endemic to human nature that it is simply *obvious* that objects exist. Such is the origin of scientific dualist thinking. But of course, the history of science has shown that, whenever a fundamental particle has been identified and named and characterised with its boundaries and habits of behaviour or decay, that particle is then found to have an inner dynamic that had not previously been recognised. The internal dynamic confounds dualism.

This human orientating habit to objectify *changing relatedness* has even been misapplied to the smallest measurable energies and distances that the Planck equations have defined.

Measuring the granularity of gravity at such small subatomic distances has tricked the language of science into potentially replicating the same object-recognition fallacy for some unwary popularisers of science, who may talk of a 'quantum of energy'. 'It', that objectified quantum, is the smallest package of energy, they say, which can behave in unexpected ways that do not fit comfortably within the orientation of space, and time. The objectified quantum of energy may stand or dangle like a false idol to the third orientating category that our neocortex imposes on heart-level systemic dynamics: the comforting experience of enduring presence. Letting go of that idol and taking hold of life instead is an emotional process of liberation. Life is all about becoming an enduring presence for others. When life is seeded within our hearts so that our thoughts and minds endure as a real presence in the wider ecologies, we no longer need any idols, or the false hopes they embody. We move on instead to a realistic and substantial hope that the life within us is ever-renewed. We are all continuous creations, and secure in the wider ecologies to which we contribute our presence in the changing relatedness of reality.

Process philosophy asserts that 'it' does not exist. It, the quantum, is a passing through. The substantial world that we know and consciously converse within is a continuous creation emerging for limited timeframes out of changes that are passing through, only some of which relationally repattern us from within. Radio waves pass through us. In many respects *it*, the substantial world, is stabilised into cyclical changes that have fractal ranges and sensitivities to other responsive change. This variability allows the dissolution and disappearance of features of the world as the process of change moves through and reshapes the patterns.

It is not even sufficient to say that atoms are aggregates of quanta of energy in probability fields of interaction. That still attributes endurance to a quantum of energy. A triquetra iconically shows a *quantum change*. The smallest *measurable* change is at the Planck length, where it is a spin into movement.

But even then, a quantum change does not *exist*. A change only gains the quality of existence when it relationally entangles with another, creating synergy. If synergic vibrational pairs were to further entangle into chains, that synergic vibration would transmit along the chain, generating energy with different wavelengths depending upon the local quantum environment. So even energy may have an internal structure of never-ending change. It (energy) moves; or, it has the potential to move other changes of relatedness patterns when informationally connected into a wider systemic network of changes. That is presence. That is why it (energy) has a deeper meaning of connectedness when seen as informenergy, increasing sensitivity and adaptability when it is substantially understood as movement. Energy connects one pattern to another, informing and reforming both in the process in a way that informs one of what is happening remotely. With a bit of reflective feedback and memory, that informational process might lead eventually to patterning within complex forms such as human beings, who can consciously understand the substantial interconnectedness of life. Then, forms can endure in memory and become values, which become a recurring presence in a person's life, filtering what they see and sense. We dream that the recurring presence of our values will be a source of joy, but it can become a nightmare for some, and may become so for many as our global environment changes. Letting go of ourselves as objects means we can more easily become a restorative presence for others in our adaptability.

But there is then another profound consequence of recognising that Planck quanta are *changes*, which do not exist on their own. We have been exploring how it is systemically logical to say that a pair of triquetral changes exists as vibrational synergy, which is relationally substantial as *something*. Each change becomes an ecology for the other. It is then also deductively logical to say that an isolated triquetral change is a non-existent change that emerges insubstantially out of *nothing*. It would not be unreasonable to conjecture that an energy wave is dynamically recruiting the synergy of paired triquetrae into extending movement. That movement may extend into nothing. In so doing, the synergy-informenergy wave may induce around itself an environment of probability, which induces new change in nothing. That change in nothing may instantaneously disperse, or it may entangle with the formed progressive wave. If this is so, then we have here a model of the continuous creation of change into substance out of nothing. This feature of triquetral quantal change might account for the measurable accelerating expansion of an open-system cosmos.

If this is a reasonable model of an open system, then the place of gravity and gravitational waves within it could not be expected to fit with an erroneously imagined closed system, and a dualist cosmos of particles and waves observed by people who imagine only the conservation of energy between them. New energy would constantly be forming. Formed energy is also degrading constantly back to isolated triquetral changes in black holes, dissipating change for further entanglement into other relational patterns, or for dissolution, depending on the non-spatial and non-temporal environment on the other side of the black hole that we cannot experience. Perhaps black holes are like cosmic compost heaps, redistributing the potential for renewal and regrowth as if fertilising the darkness. That would fit with process philosophy in one integrated triquetral cosmos moving with informenergy.

Letting go of religious or philosophical dualist language

Having painted a picture of an informenergy cosmos, *process triunity* could equally well account for the inner ecology of a person's continuous creation, renewal, healing, and diversification, embedded as the person is systemically in the whole. In Part

III we shall look closely at the inner mental processes that can degrade that systemic connection. By disrupting the triquetral balance in the inner ecology, people can become separated from the land, isolated from each other, and internally dissociated, veiling the process informenergy within layers of dualistic misunderstandings and inner heart-level brokenness. More than 3000 years of intelligent reflective thinking about the experiences of life that individuals report as stories has led to systems of ideas that philosophers and theologians protect with emotive energy as if their personal identities depended upon them. Disagreement overrules curiosity about the inherent diversity of life. People erect emotive walls and barriers to others, rather than responding adaptively to them to explore the restoration of triquetral balance in their lives.

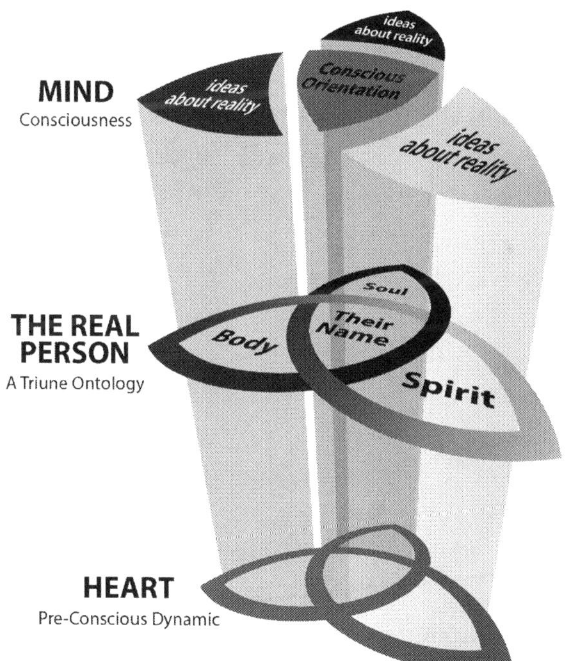

Figure 17: Mind and heart qualities of a substantial triune dynamic

Figure 17 may be helpful to bring together much that has already been said about personal identity. The single triquetra in the centre should be imagined extending from its yellow-coloured point (pale in greyscale) into a double triquetra, as in Figure 5, which is the core icon of the triquetral model of personal development. This imagined double triquetra shows the synergic modulation of personal values by interaction with their physical and social ecologies. A person is their values. Their values are their anticipations. Disordered or destructive anticipations generate instability in the wider ecologies of life. Valuing order can also connect people into stable domains as communicating and adaptable groups, as shown in Figure 9 in Chapter 3.

To talk about the inner nature of a person is difficult, however, when language above the orientation horizon objectifies the process behind lived experience, a process of emergence and dispersal. Orientation tries to place objects into timeframes and spatial orders that not all people agree on. Named 'parts' of the processing person, when referred to with the pronoun 'it', give a spurious sense of hierarchical power to the thinking mind, as if that mind is somehow looking upon 'the person', who is now mentally dissected into parts and named to death.

The way of integration that is an alternative choice to this analytic frame of mind is to tell the story or narrative of life as seen in the unique perspective from this personal place in the whole. The story told is a person's best attempt to summarise and integrate the unique diversity of their vast range of relational experiences and changing hopes, pleasant or unpleasant. But no story can give a full picture of the whole, which needs to be explored with curiosity, experienced, and discovered continuously, while active and equally while stative at rest. In that way, both substance and understanding can be integrated in a mindful and restful peace while walking and talking with others. The whole is still there when we awake, because it is a process of continuous creation. We are each one relational frame within the whole, who communicates and resonates with others. We re-emerge into life as we awake to our history, future, and presence now.

The person's name

The person who relationally tells or shares their story of exploration and discovery has a name. That name is the author of their evolving story. The author's name therefore has 'authority', which means only that there is a personal source of the changes that radiate from the telling, or living, of that story. The changes experienced by others are *not* merely random. They are being informationally shaped by the choices that a person has remotely made, the author. If people feel an inner ecology pressure then to submit their own choices to their experience of that remote informational change, they may misattribute to that word 'authority' the connotation of a hierarchical order and a power structure. This attribution of oppressive power to another's authority may be motivated by fear, which is the anticipation of grief rather than of joy at the experience of diversity in another's story. Grief and joy are modes of the enduring relatedness of the whole, called love. Triquetral imbalance due to misunderstanding another's motives can lead to imbalance of love, and eventually to brokenness and separation of relatedness. Hearing the author's name may unfortunately put love into its grief mode through misunderstanding the story, rather than its joy mode. In a triquetral cosmology, equality of value among diverse perspectives on life is foundational to maintaining life's integrated movement.

Hierarchy, therefore, is not inherent to the triquetral interpretation of authority. Author and listener have equal value in the unity of stable domains, but diversity of roles and knowledge to share informationally. Equality of responsiveness opens the systemic process of resonance and coherence, which energises adaptability and joy. Diverse people become ecologies for each other as substates of the living whole. The storyteller is a name actively shared. The story heard is a stative potential for change from an author who seeks relational repatterning. The story can be informationally reactivated whenever it is reread, or listened to, or viewed again, so that feedback induces learning within an adaptive system. Stable domains result when diversity comes into unity, even remotely and across time, whenever the experience is heartfelt.

Authoring stories shares informenergy. Informenergy coheres stability *from within,* at a heart level of reception and responsiveness. Triquetral stability in adaptable living does not arise from an external source demanding only conscious assent to submit. It emerges from the coherence of love within. Authority being misinterpreted as power and control may have followed misunderstanding the intention of a storyteller to lovingly induce change of one harmful feature of a listener's values, the listener then generalising that misunderstanding to the whole of life. Storytelling informationally liberates people to learn from diversity. Stories could be told in a song, or a work of art, or a dance, which are all more at a heart level before words complicate the process. The whole physical ecology of the cosmos is telling a story in movement. Hearing that story fills out the thinness of an isolated life. Listening to the heart of a story can restore equality and joy after grief, forming stronger personal identity to move on into a richer life.

People write their names across the world by their behaviour and words. They leave traces of their signature inscribed in the physical and social environments touched by their lives. That signature trace is informenergy primed and shaped by personal values embodied physically in those core brain filters. Mind and matter meet in the dynamics there. We see the mark of people's lives in their homes, collections, workplace impact, memorials, and in many other embodied ways. A person's name, when spoken out, is triquetrally an embodied, substantial summary of their story, which may be understood, and unfortunately misunderstood.

Classical terms applied to personhood

The intention here is to refresh the value of previous systems of cultured thought and behaviour by transfiguring parts into perspectives. Figure 17 uses the classical terms 'spirit' and 'soul' alongside 'body' to describe diverse perspectives on the intra-active movements of a self-orientating person. The classic terms can gain new *relational frames* of meaning that successfully blend them into a picture of a relational person moving by choice with inner diversity-in-unity, in whom dualist thinking may have inserted boundaries that potentially harden the heart. There is, however, much of lasting value to be understood and retained from the great philosophy and religious teachers, who may have been process thinkers while their commentators and their students inserted the boundaries in their attempts to analyse and discuss what had been said about life, rather than just living it. The new relational frames can also connect these perspectives seamlessly into the new physical sciences, especially now that most quantum physicists have had to become process philosophers as well. Triquetral cosmology is pre-religious. It is not anti-religious, nor is it anti-science. It is an adaptive systems approach to understanding personal growth in a quantum gravity, open-system cosmos. It opens a way to think clearly about survivability and the endurance of compassionate life by choice during times of change. The dynamic systems view of informenergy, as both material and personal connection, may nevertheless be a source of awe and wonder for many people, unveiling the potential for life's renewal in the resonant movement that is physical matter. Understanding the substance of movement may initiate a dance or a song in the heart, which reawakens a sense of the sacred in awe and in science as a lasting value to enjoy.

Soul

Tables 3.1 and 3.2 in Chapter 3 show the language equivalents in different specialist settings of the principles of organisation that also move persons in relation to each other. In a triquetral cosmology, there is nothing static about life. Movement within emerges into patterned forms that transfigure life's changing relationships. This inner process becomes recognisable at the human level of experience in growth, new beginnings, healing, changes of heart, renewing of the mind, and restoration of orderly responsiveness after times of chaos, or destruction, or loss of whatever is valued.

Soul may be thought of in this context as the inner life of a person, where complexity generates an inner poise, an inner ecology stative state that could critically change into unexpected courses of action when environmental change enters those equations of inner balance. The notion of soul (Greek, *psyche*) could include functioning, harmony, change, and turnover, all shown in Table 3.1, and all filling out the picture of the life of a person. That life is in heart and mind modes embodied in all three ecologies of the person simultaneously. The inner life cannot be separated away from its embodiment and its relational dynamics locally, and remotely also, which may be seen as the spirituality of the soul in the integrated heart level of life. To picture soul as a boundaried noun, '*the* soul', requires a dualistic separation of life from energy-matter. Once separated, how can it ever reconnect? Or how can it move matter? The physiology of matter in the body is moved, however, by our relatedness to changes in our environments. In a triquetral cosmology, matter is moving with the same informenergy that moves the inner heart of a person's soul-life. At heart we are energy that matters. And that energy has an inner structure of synergic relatedness.

The noun concept of soul-life has been present from ancient Greek times onwards (and before), in the way people have discussed the possible continuation of life after death of the body. It was a Platonic norm to consider that the material world is corrupted from perfection. This could leave people without hope unless the person's life is also considered to be separable from the world in some way, as a noun, self-existent in itself with the potential for perfection. Many people do aspire to purity and healing into wholeness (although not all). To have a psychological boundary allows protection mentally from corruption. However, that mental separation into a self-preserving noun is itself a corruption of *process triunity*. The mental separation of the psyche from intra-active agency in the world can harden the heart into a dispassionate state rather than a compassionate one, or can separate compassion mentally from effective action in the wider physical and social ecologies. The soul would then become a self-centred ego state, which in triquetral cosmology it is not. In process triunity, the ego is an inner ecology state of balancing by choices made in all three ecologies, which are the context for action or rest within a transforming cosmos.

In a triquetral cosmology the soul of a person, personal life, can be resonantly synergised across time and space if sensitive and attentive. The whole person is resonating triquetrally with the presence of others. That soul-life can awaken in response to a call by name, and emerge into context, *and disperse*, and perhaps re-emerge relationally from pre-conscious states of rest. The dispersal of the life is a vital feature of process. The endurance of soul is in the memory traces that the person's life signature has written into continuous creation. The enduring soul is not self-existent. Within this new triquetral order, soul *is* the informenergy of feedback in communications. This *feedback* synergises the potential for an emergence of renewed life.

A similar psychological process separates an atom as a noun from the rest of the cosmos. It is only boundaried by human choice to focus that way. The ancient Egyptians had a similar psychological boundary, seeing the eternal souls of pharaohs as stars in the night sky, separated from and far above the rest of the population.

Atoms and molecules that were thought at one time to be discrete from each other, and separable, may now be thought of as having their own physiology, in which exchanging quanta emerge as patterns at different environmental scales, which affects light, and gravity, and thermal vibration all at the same time in our own experience as living soul. Life emerges synergistically at every level in a process of fusion. The soul of the cosmos would then be at that level of the enduring warmth of its fusion, and equally in its dispersal to fertilise the darkness with light. But soul does not exist on its own. It is not an ego state in isolation. So, the cosmos cannot *have* a soul.

Spirit

Likewise, the classical term 'spirit' (Greek, *pneuma*) may be mentally pictured in a triquetral cosmology to include all of the terms mentioned in Tables 3.1 and 3.2, namely social, cause-effect, output, spirituality, relationships, radiance. In maths and engineering physics that notion of relationality is seen in the equalising balance of an equation that describes a transformational process, or connects the impact of ecologies on each other, such as when building a bridge. To restate this process in a triquetral perspective, spirit is the name given to that feature of movement which *connects synergistically*. Spirit is that feature of movement named relatedness.

The Greek term for spirit, *pneuma,* literally means 'wind'.[1] In the ancient world, nobody knew where the wind came from or went to. Its power was known by the effects of distant change on the local forms of life, such as windmills and sailing ships, or destructively in storms. This paints a mental picture of local connection to a remote source of change, beyond a horizon, or beyond an individual. If that idea of connection is only externalised, however, and projected out from the subject as a source of power beyond 'it' or 'me', then a boundaried-noun mode of thinking can arise about ancestral or other spirits 'out there'.

In a triquetral view of life, a boundaried spirit would therefore be a disembodied relationship. That weird idea generates fear, in anticipatory grief about the potential loss of one's own personal agency. It is a fearful and disorientating thing to anticipate being disempowered by having one's own choices limited by *the* unknown powers. It can turn spiritism into a sort of paranoia, harbouring a victim mentality in the heart that may be totally unjustified. A thinker imagining alien powers and spirits is primed to try to appease them behaviourally to reduce the emotional corona within of anticipatory grief. There is now a triquetral alternative. It is to prefer *by choice* to be primed by harmonising with enduring life. Understanding how the resonance of feedback is also at the heart of substance can improve the power of choice to explore one's own place from which to influence wider life. Resonance establishes a true personal agency, which by choice can keep life in balance.

Rather than abandoning the notion of spirituality, as some popularisers of science have asserted to influence society and reduce their anticipatory grief, a triquetral approach to belonging in society in a cosmos that generates and renews life can help to *remove* fcar of the unknown. Triquetral diversity means that, within *any* mental framework that makes temporary sense of the cosmos, choices of behaviour can still be made that balance reason with values-based insight and emotional awareness. Systemic integration is foundational to security. Inner separation through fear creates unstable inner states. By seeking systemic integration, a person's spirituality can evolve and mature over time as more is understood about life. As personal security increases with a sense of belonging in the universe, the focus can grow less on guarding against fear, and more on the protective impact of the presence of loving kindness and compassion.

Body

In any semi-stable environment, from dark, deep seas to boiling geysers, and from rocky mountains to fertile plains, the physical universe has within itself the potential for life-forms to emerge and develop, and for the seeds for life to self-replicate across time. The embodiment of movement in a living form is its physiology, its life chemistry. That physiology of embodiment is the turnover and exchange of informenergy between body and physical environment.

Embodiment and environment are inseparable. Their enduring connection in movement is informenergy at every level of life. If a form gathers in one place, then it affects the probabilities of forms gathering anywhere and everywhere else, to a greater or lesser extent. The substantial material energy accumulating and held for a timeframe in a living body is a product of the biochemical responsiveness of DNA and RNA to its local environment, in cell nuclei, mitochondria, bacteria, and viruses. As mentioned before, genetic material is a form of memory of survivable living forms (phenotypes) in changing ecologies. That living form is not a static object with clear boundaries, as the temporal lobe sensory association area would have us believe. It has slight variations of fractal range in different spatial environments, at various times of day or night, and in responsive presence at the vast range of events in which this body is intra-actively present. The pattern-recognising capacity of the neocortex and immune system working together can detect much of that fractal range and locate that recognisable body in mind and heart as a person or pet animal with a name, but there will always be some surprises after a considerable time lag, when people will experience surprise at a noticeable change and have to readjust their memorised range of 'normality' for this named person.

The Enlightenment Science focus on a primacy of forms, and on developing technologies that measure changes of relatedness between boundaried forms, has resulted in people naturally thinking of physiology in terms of the biochemistry of molecules. The chemicals of life dominate thinking, which affects planning and anticipations and valuing. Chemicals and biochemicals are the dust of life. Dust needs a bit of water and lightning to add movement to *it* within primordial mud.

However, the New Science is a paradigm shift from this view. Now, change of relatedness has primacy over form (although actually all three are in perfect balance). The inclusion of relatedness in ordered chemistry is now needed to understand the dynamic substance of forms. This paradigm shift in thinking and valuing means that the notion of embodiment as a dynamic should now have primacy over the notion of a boundaried body. The separateness of body, soul, and spirit becomes meaningless in such a dynamic.

A triquetral cosmology framed in this New Science replaces matter's solidity with a notion of substance as *informational resonance* between a minimum of two interrelating changes.

Informational resonance can scale with fractal ranges to any and every level of order and movement in the cosmos. As previously mentioned, the *molecular physiology* of biochemistry is a relational exchange between the embodied life-form and its environment. In exactly the same principled way, there is a *quantum physiology* of biophysics that is a relational exchange at atomic and molecular levels; and there is a *social physiology* of bioinformation (verb) that is a relational exchange in society and families emergent mostly as emotion. The elements of the adaptive system are different, but the principles of adaptability and responsive movement are identical. This is informenergy generating resonant life.

Resonance requires turnover and feedback cycles, out of which self-regulation can emerge, bringing stative stability with a capacity for critical change into active movement. Stative, cyclical stability is a form of memory in adaptive systems. Stative states bridge time. In living systems, the important feature of time is not, in fact, its linear capacity for cause and related effect. It is *timeframes* in which whole nested systems can change and grow or disperse, so that new beginnings, healing, and restoration can differentiate one state at the end of any timeframe from the patterned state that existed at its beginning. This is embodiment, as contrasted with body.

The mathematics of physics works in both directions in formulae for changes of relatedness between variables. In biophysics, this bidirectionality of quantum physiology could also be true, as contrasted for example with engineering to build bridges that do *not* have critical changes. Within timeframes, matter at a biochemical level (as studied so far) seems to progress in one direction only, but when an informenergy view of the embodiment of changing relatedness is placed within a timeframe, within any timeframe, something different emerges. Within a dynamic timeframed living ecology, informenergy integrates understanding and substance. This introduces *final causes*[2] and design features to personal physiology. The body no longer moves just past 'efficient' and 'material' causes. When a heart-mind level of understanding establishes a timeframe to compare change in patterns of relatedness, the anticipated pattern that is valued becomes a final cause of the reactivity of the person. The anticipation pattern shapes life in a way that is before time, or beyond space. The same would be true for any living form that can anticipate and has choice agency.

Enlightenment Science specifically and intentionally excluded final causes from its experiments. It allowed only the 'arrow of time' from past causes to measured effects, and in so doing killed the embodiment of living informenergy, reducing living movement to boundaried objects as parts of the body under the control of the analytical mind.

Understanding movement adds final causes into substance. Final causes are the higher-order aims, anticipations, and values that contribute to movements of the system as a whole. Within consciously chosen timeframes, the power of choice works to bring understanding and substance together. In a person's experience of life, this meeting creates a place in the physical world for the inner human heart, where embedded values are final causes for intra-action, and the analytical mind adds a design element to those values.

Mind and heart as modes of the person

Figure 17 shows the inner ecology in mind and heart modes. The personal ontology, the real person, is the informational resonance of feedback loops connecting this unique locus of inner feedback into wider feedback systems.

We have already differentiated awareness, which characterises social feedback sensitivities at heart level, from consciousness, which adds comparative feedback of analyses from the neocortex into heart-level processes. 'Conscious awareness' is thus a reasonable term that describes the integration of these patterns of informenergy in the *rationalising heart* of the personal ontology. There is nevertheless substantial value in differentiating its awareness and conscious features in terms of the differing feedback loops, but not in terms of imagined ontological boundaries between them. They are equally vital processes interacting.

Awareness is the ecological engagement that, in process terms, is best described as mindfulness. Much Greek philosophical thinking debated a concept of *nous*, however. This is commonly translated into English as 'mind' or 'intellect', which is the capacity for rational thought, perhaps even the conscious containment of the reasoning faculty along with will and intuition. *Nous* omits the heart-level feature of emotion, however, which triquetral cosmology integrates into mindfulness as preverbal evidence of the ecological basis of personal values that move thought. I mention this here to avoid confusion later when writing in Part III about the distortions of mind that can influence heart-level self-integration. Another translation into English for *nous* is 'awareness', which in triquetral cosmology is the term reserved for heart-level sensitivities. The Stoics believed that sensitivities were corporeal, setting themselves against the Platonic belief that they arrived in life experience in a numinous or spiritually esoteric way. This Stoic interpretation of mind is the meaning adopted in Westernised psychology. Mind in Westernised psychology is the rationalising refinement of raw corporeal sensitivities.

This interpretation was assumed by psychologists, however, before the paradigm shift in rational thinking arising from the quantum revolution and process philosophy. Now, even raw corporeal sensitivities are being considered to have more than only anatomical and biomolecular 'sense organs' as their source. Biophysics is opening new ways to broaden our picture of personal sensitivities that are not boundaried in the same way that materialist or positivist thinkers have tended to picture the molecular body. These informational resonant sensitivities move the inner heart with more subtle patterns, as intuitions. They require interpretation and orientation by rationalising feedback into the heart from the neocortex. However, intuitions do not need to be pictured as spiritual in the sense of an esoteric energy set over against material energy. These heart-level movements and intuitions, which can break through the orientation framework of space and time, are simply more remotely resonant and patterned feedback loops of the same informenergy that shapes physical matter.

The inclination of the human heart

The inner heart moves people to engage in the world in ways that are primed by the filtering process that has been initiated by neocortical rationalising. It thus has *an inclination*, which is a predisposition to react in some internally valued way. There is a meeting point here with Stoic thinking, in the notion of the inclination of the human heart. The Stoic *nous* was considered to be soul 'somehow disposed' to rational thought. That is perhaps the same meaning as the inner heart being inclined to think, reason, or react in certain dispositions.

Self-awareness of that inclination is difficult to achieve in everyday living, however, arising as it does from unnamed values that are action-orientated. Awareness of that personal inclination is commonly lost to insight for people who also 'live in their heads'. Living mindfully instead, by which I mean in embodied movement and responsive social presencing that includes the informational content of emotion, reintegrates the heart and mind modes of being.

By wilfully enlarging heart-level sensitivity to wider systemic influences, values can evolve if curiosity allows mismatches through the filters to be magnified for consideration in the conscious mind rather than screened out as distractions. People may impact societies more effectively when their values can mature as their understanding of life-enhancing options grows. For that, a gradually clearing picture of life as systemic process is vital. Then the mental veils that hide the beauty of systemic and resonant reality can lift.

The inner heart is informationally like the core of any fruit. It is the inner source of self-organisation, the source of growth and ripening or maturing. Growth is a gradual process of gathering available elements little by little in a physical environment, in an ecology, around the seed of life that is bearing fruit and creating more seeds in that process. That fruit reconnects into its environment cyclically. Ultimately this includes a death process that seeds the potential for new life, not a death process that is the end of systemic life. Human values leave a legacy, and we receive legacy through our shared ecologies.

Ancient anatomists and theologians imagined that the physical pumping heart might have had such a core role of feeding growth of the body. Descartes thought that the pineal gland in the brain might be another source of a dualistically different mental homunculus. However, rather than trying to locate the source of life in any one physical place, *triune principles of organisation* locate that source of life everywhere, in the entanglements of triquetral changes and as the release of informenergy by these fusion changes. When the dispersed potential for life's renewal is understood triquetrally, and pictured extensively, the social-environmental feedback loops that extend from and return to a person's inner heart may develop a person's sensitivity in a way that inclines cerebral thoughts differently, not to filter out the unexpected and indeterminate possibilities, but instead to willingly open new awareness of the life context for choice options.

Hardening the heart and softening the heart

As a complex adaptive system, the inner human heart is constantly in a changing state of integration or disintegration. It may be at peace, and it may be troubled, but these equally are dynamic states that can bifurcate into vastly different states when pressured or released by changing informenergy inputs.

The inclination of the heart to shape thoughts and activity can rapidly change in association with this changing state. The emotional corona around its embedded values rapidly initiates changes of state as predispositions to engage with or disengage from situations. This is occurring continuously before neocortical activation generates orientated options of response to those reactive situations, adding the capacity for choice. This timing of predisposition changes can be measured in neurophysiological studies, leading some scientists to question if there truly is free will, because they are stuck in a mistaken belief that the choosing person is 'located' in the neocortex. Hopefully, the way is clear by now to see the nature of this misunderstanding. It arises by thinking only in terms of linear time, and omitting timeframes of predispositions from the core-cortical-core feedback loop. The predispositions arise from anticipated values that have already been embedded in the core brain, prior to the social feedback loop inputs that generate the emotive corona and curiosity. That resulting corona requires the freewill choice to update the filters with renewed anticipation values. Timeframing is more important than time-sequencing to understand the living human person.

The timeframed feedback that alters the state of the inner heart can create a disposition of the inner ecology knowable as hardening of the heart, or softening of the heart. Hardening of the heart is a dynamic state of increasing the density of informenergy patterning in a way that preserves a particular set of personal or group or cultural values. Softening of the heart is a dynamic state of increasing the receptivity and sensitivity to the informenergy input of social feedback loops in a way that opens curiosity and responsive adaptability in social settings. Both can be responses to a perceived injustice.

Hardening of the heart may increase the deterministic probability of an *external* change in response to injustice despite risk of damage to the self. Softening of the heart may increase the indeterminate potential for an *inner* change in response to injustice, for example a merciful engagement with suffering to bring relief. Unfortunately, both of these heart-level states can also develop for self-centred reasons, not focused on injustice, but perhaps on greed or fear, pride or shame, and disruptive pleasure-seeking or disgust. There is an infinite spectrum of personal motives and reasons for hardening or softening of the heart. Hardening the inner heart makes people behave more like particles that collide at their surfaces, while softening the inner heart makes people behave more like waves that transform situations from within. The informenergy view of triquetral movement shows how there is only a fuzzy boundary between inner state and surface state, not a hard boundary. The degree of fuzziness is the reality of relatedness in the movements of changing forms of life. There is only the personal choice of values, which pre-sets responsiveness. The consequential degree of fuzziness pre-sets a process of responsiveness that is knowable as *indeterminism*, and unpredictability in behaviour.

The last point to make about choosing to harden or soften one's inner heart is that both are equally courageous. Both are equally brave choices. They are not in themselves manly or womanly, as some culturally classical thought had formerly differentiated. There is a profound equality in this core diversity that balances within and among all relating persons if we are to be truly human from deeply within.

A broken heart and healing of the heart

A true cooperation between heart and mind embeds the person in a social reality that is mostly intuited. We guess our way through life, and often need to seek affirmations. Dislocation of the mental system of organisation from the social one, however, may distort the potential creativity that arrives through social feedback above and below the orientation horizon.

A mental informenergy system within the cortico-cortical feedback loops can generate imaginations, daydreams, and dreams that are neither orientated in space, time, and social reality, nor derived from real-time sensory inputs. Such imaginations can turn into vain hopes, obsessions, delusions, and hallucinations. They can leave people vulnerable to manipulation by others of wounding intent. Vulnerability arises from lack of insight into how to embed one's own personal life-enhancing values into one's own inner heart. Then others who know how to seed theirs into the hearts of others can do so, perhaps to gain power over them.

Imagination may, of course, also add great beauty, hope, and inspiration to cultures. A life-celebrating imagination can artistically diversify life, seeding new sensitivities and movements into lines of thought and emotional coronas. But when, on the other hand, a mental mode of living distorts or oppresses the core values of the social brain, fractal scaling into substance may sadly corrupt and undermine society.

One method of manipulation or bullying is to modify the names that alert a person's heart and awaken them to renewed life. To honour someone who has been outstanding, people might add 'The Great' to their name, thus extending the 'trace' in history to those who read or hear about this signature life later. More sadly, the same is true if people want to caricature someone they dislike by 'calling them names', to reduce their trace in the neighbourhood. However, inspiring people to creative arts and crafts and dance is one way to restore the storytelling heart of a person to reshape their life. Through movement that stabilises their inner ecology stories in clear timeframes, people become more inclined to reconnect into mutual resonance with others, which is the healing process.

Another method of manipulation is through distorting the body image that someone constructs, adding doubt and hesitancy when that body image speaks of a person's changing qualities of relatedness. Disliking body image leads to internal dissociation, unhelpful anticipations, and distortion of values. Body image is a cognitive veil that covers heart-level synergies with anticipations. If that body image includes isolation or hurt or disappointment rather than connectedness, then behavioural problems may follow. If an improved understanding and naming of personal values can be heard by an emotionally available adult, then relational responsiveness around named values may start to restore healthy adaptability and systemic wholeness, altering body image secondarily.

As people become more life-aware, the synergy of restorative relatedness is accessible everywhere. If brave to enquire, a conversation could spark enough synergy to enlighten new paths that, in time, may restore hope and deepen relationships. People need to call on each other's name to attract attention, so that transformation can follow. Even those people who are hurt can call on a name for help, changing the story.

Healing is a synergistic process. There must be a wider system into which a healing transformation integrates. Self-healing is dissociative if it does not lead to a triquetral balance that includes compassionate action in qualities of relatedness. Small compassionate acts of kindness are clearly recognisable as systemic synergy. Valuing that synergy transforms life peacefully. These compassionate acts may be concealed, so that they work dynamically at a heart level and may become known only later by intuition. The key point is that healing synergy at this level is not self-centred. It is *double triquetral*. It balances all three principles of organisation of life—form, change, and relatedness—into a conversational ongoing process. No matter how knowledgeable an individual becomes about various aspects of life, unless that awareness turns into a conversational or communicated exchange in some way it remains aside from synergic reality. Self is a vital component of synergic life, not a centre.

The healed person is an author of value who receives in humility from the wider whole and contributes also their unique balance of life-enhancing values as a participator in the wider life.

The thoughts of your heart

The light of conscious awareness is sensitivities remembered. Sensitivity to local environments is important for personal development and the growth of kindness. It can be darkened when the memories hurt, and there is no 'emotionally available other' to relate to and heal with. It can turn to utter darkness when insensitivity and isolation team up with lack of remembrance. But the synergic capacity to heal remains there when making life-enhancing choices, potentially restoring light and hope.

A mindful balance of the personal values that are filtering sensory inputs can become critically unbalanced if people narrow their sensitivities and cease being reflective. We have seen how the patterning of emotional preparations can be mapped in relation to a life situation, either an anticipated one or a remembered one. When practised, the mapping can also continue in mind during an ongoing situation. This process slows down time, creates even a gap in time, and creates options for timeframing responses. It makes explicit the implicit inclination of the heart to respond. It reveals the emotive patterning that shapes the thoughts of your heart. It is a cyclical process, which may trap people in unhelpful habits of thinking, behaviour, and emotional processing, but its slight variability has the potential also to evolve fractally into something different and more openly integrated.

The core social brain is thus a focus for personal unification in a way that allows infinite diversification of personality, character, and spirituality. The processes allow disintegration into chaos, and equally allow healing and renewal of implicate order from that chaos.

The inner heart of a person filters their thoughts. And thoughts adjust the informational filters. Indeterminism is built into living physical systems, which are poised in a critically dynamic balance, allowing adaptability. In the developing social human being throughout their lifespan, the indeterminism of their physical systems becomes stabilised by neural and hormonal feedback. Within this, a person truly does choose how to shape the physical cosmos locally to some extent, and in relationship with others. Some people may disengage their responsiveness from that responsibility; but a change of values alters the thoughts of all our hearts.

Just as fusion energy is safer and cleaner than fission in the physical ecology, so the same is true in the interpersonal one. In the internal ecology, fusion requires the same sort of feedback containment as in a nuclear fusion reactor. That containment relies on having the insight to name personal values. This sustains the diversification of order, breeding character by empowering choices in a changing world. This is true both for managing a power generator and for managing a human society. Explosive disintegration can occur at every level of life. However, the slow process of compassionate rebuilding can restart after wounding, because triune principles of organisation are informenergy in both the quantum source of every particle, and the triquetral source of every person.

Part III

Conversations between People Who See Life Differently

Chapter 8

Three Analytical Perspectives Can Disrupt Conversations

Preparing to adapt in conversations

Mutual understanding between people can break down. People may seem to be in different worlds, with such different values that no heart-level connection seems possible. Hurt, disappointment, and confusion may follow. One way to possibly make sense of this may emerge by seeing how vastly different worldviews can result when a shared heart-level conversational orientation twists as people start thinking *about* life rather than living it.

In this chapter we shall see how three very different *analytical perspectives* on life can result when people take a mental step back for reflection. Although this increases the informational content of thought, it inevitably breaks conversational orientation for that analytical timeframe. The resulting worldviews are so different, however, and the logical deductions in each so different, that a sense of rightness and wrongness can emerge rather than a delight at the diversity that can be seen inherent to life. Misunderstanding can lead to separation and relational brokenness. Understanding instead how they all derive from the same shared triquetral informenergy seen in diversifying perspectives may help people to maintain conversational connection. Maintaining self-confidence with diversity can give a direction to explore and grow in understanding.

This Part of the book aims to develop the mental flexibility to see life equally well in all three diverse primary perspectives. If people are interested only in pursuing one single perspective on life, a mental habit of thinking analytically may build up internally, restricting the inner heart, which potentially breaks fusion and pushes logical deductions to an unhelpful and self-isolating extreme.

The three diverse 'primary analytical perspectives' focus attention on the *changes* affecting situations, on the *qualities of relatedness* that seem to be active within situations, and on the *forms* of life that are noticeable among triquetral informenergy patterns of changing relatedness. Just as an artist can mix a full colour spectrum from three primary colour pigments, an infinite spectrum of secondary worldviews can result from mixing the ways that life seems different in each primary analytical perspective. Habitually seeing life through the lens of only one analytical perspective may be problematic.

When temporarily or habitually adopting an analytical perspective, a thinking individual creates an I-it distinction of self from lived experience. Martin Buber, in his paradigmatic 1923 book *Ich und Du*,[1] described how this analytical mental shift necessarily disrupts the I-thou (or *me and you*) mode of personal identity, which is here called *conversational orientation*. The triune model of personal identity clarifies how the I-it mode he describes has within it three I-it perspectives. These have a predictable impact in different ways on beliefs and behaviour. The evidence for this is that these patterns of thought, belief, and behaviour are already seen globally in cultural and religious diversity, some of which lead to conflict as people sadly break apart into factions. Anticipated grief at the loss of 'being right' if the other perspective is shown to be equally justified may be emotionally damaging for some who feel vulnerable. However, if people can trust the underlying triquetral platform of informenergy that unites all three analytical perspectives, then a secure grip on each other's beliefs can move life forward constructively together.

Grief for loss of personal identity hides in ideas

Personal identity is integrated in the core brain, including its limbic system, where self-awareness arises from anticipating the impact in the world of chosen action or safe rest. Anticipation patterns are embedded physically there as personal values, bringing an emotively patterned affective physical response; biochemistry is the other half of personal identity, spread throughout the body as the immune system's memorising and pattern-recognising capacity that also affects responses. Feedback movement therefore underlies personal identity.

It therefore follows that ideas and physical emotions are intimately connected via personal values. They are not truly separable without life becoming in some way less valued. If ideas get challenged that have any connection with personal identity, those ideas may be defended intellectually, but with an emotional intensity that amounts to social survival. That is the academic life of conflict as much as it is the political or religious one. The cyclical dynamic that can entrench people in a habitual single analytical perspective starts when its associated ideas become personal identity values.

In each of these analytical perspectives, the resulting 'it' of I-it may seem to be a different type of *essence from life*. 'Essence' comes from the Old French/Latin word for *to be*. One or two of these potential three 'essences' may feel more valued than the others—matter, information, and a vitalist spirit. In relation to these perceived potential essences, the 'I', the 'me', also reduces from the conversational wholeness of informenergy, as we shall go on to see how. Conflicting attitudes in politics, culture, science, and religions can be mapped on to these analytical perspectives and reduced personal identities. Anticipated grief is then wrapped up inside the ideas generated in these analytical perspectives, because that perspective becomes an attractor state with simplified boundaries, within which a sense of personal security and 'rightness' feels comfortable.

Logically, with both a systemic and a deductive logic, all three analytical perspectives are *equally valid* as mental reductions to mindfully view any feature of life, in a *me-and-you* conversational reality. Each reduces the triquetral assumption base for thought to just two of the primary features for orientation in life. This narrowed analytical platform provides the base from which to gain a perspective on the third feature. The fully balanced conversational orientation temporarily but necessarily disrupts into a critical change, or a bifurcation, of the mental system into a different state. If this state feels safer, because life seems more predictable, the attractor state may become difficult to shift back into fully triquetral balance, where diversity is naturally generated. To transfigure this core level of personal security and restore respect for diverse personal identity, insight needs to grow into why diversity is needed for adaptable living in the face of change. Picturing the double triquetra as an icon of conversational security might be the stable platform to insert a healthy gap in time from instant reactivity. That gap is, in fact, a timeframe inserted into a linear time sequence of responses, in which timeframe a more life-enhancing reflection may be made on choices to respond.

Developing the mental agility to move freely between all three perspectives on life will produce a rich understanding of the healthy unpredictability that preserves life in changing environments. Many people *are* able to do this quite spontaneously and naturally, but most may discover this is a new mental skill that needs practising.

In the same way that an artist may mix three primary colours on his or her palette to create vibrancy in our visual worlds, so also an individual thinking about life could mix and match these three primary analytical perspectives and their associated essences to create an infinite diversity of worldviews of *what life is all about*. And just as people have likes and dislikes in the visual arts, so also people will agree and disagree about the way the world works when they compare their views. However, underneath that vibrancy of conversation can be a triquetral peace of profoundly synergic relatedness. There may even grow a quiet, inner joy on discovering that the other person, whose ideas had once seemed so odd, shares perhaps 90% of your own named personal values.

So, how to approach this chapter, where readers are sure to disagree with some parts, and enthusiastically give their assent to others? The key word has to be *curiosity*. Encountering something you disagree with, or have doubts about, or find challenging, needs a response here that says, "I am ready to learn!" Suspend final judgement until the end of the book. There may yet be far greater levels of security in life that emerge. We can be motivated by grief emotions to learn and explore after disappointments, provided we understand grief as a mode of exploration on a creative journey to reconnect with others. All of this takes time, however, and testing, gently and conversationally if relationships are to grow with a self-renewing inner stability. We have the power of choice, to come through stronger together, or to withdraw into self-defence.

I-it modes of thinking are the mental tricks mentioned in the subtitle of this book. They veil the reality of social and ecological synergy. The capacity to benefit sustainably from seeing the world in a newly balanced way requires triune vision. Exploring that way of conscious awareness, thought, will, and relationally moved emotion may require a short phase of grief while questioning previously treasured ideas. Character formation and improved social connection can emerge, so that violence is no longer needed, as trust in a bigger dynamic system fills us from within and restores personal security and wholeness amid its diversity. When people start to converse instead as equals with those who differ from them, not aiming only to profit from them, then all three previously divisive analytical 'I-its' will begin to make more sense together.

If this hypothesis is proved true in your experience, and there are three primary analytical perspectives to become familiar with, then for every 'yes' as you encounter one that you feel inclined to give your assent to, there may be an echo of 'no-no' from those readers who prefer the other two. Somewhere in the middle is the conversational mystery of truth about personal life, which is revealed through conversations with those who differ from us. If loss emotions are a problem, the Emotional Logic Centre can offer guidance in how to harness them into constructive exploration. Let the exploration begin.

The individual—potentially an isolated person

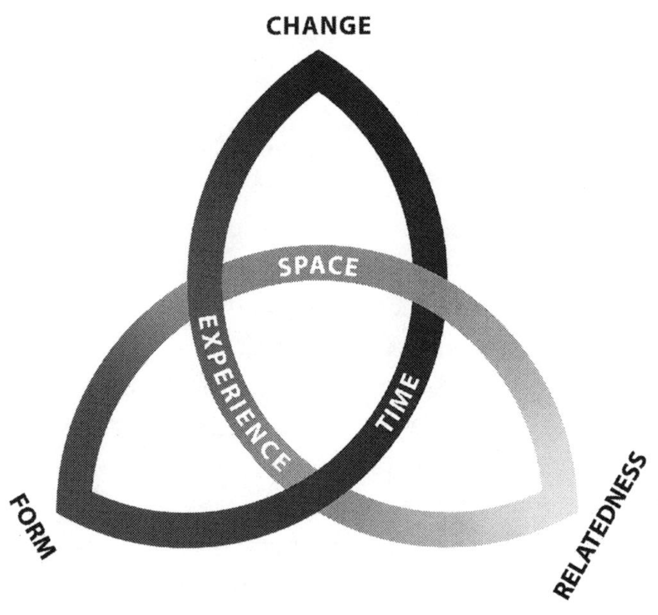

Figure 3: Three primary constructs for orientation

Figure 3 is replicated here to ease comparison with the three analytical distortions of it that will follow. It shows how an individual may informationally construct an orientating mindframe of space, time, and experiential presence. The four potential phase spaces of Figure 3 could be named with the classical terms shown in Figure 17: Mind and heart qualities of a substantial triune dynamic. They identify the key features of movement as they become present in personal life. (Please note that in Figures 3 and 17 there is a reversal of the relatedness and form corners. This may emphasise how important diversity among people is for conversational synergy. The colour and greyscale coding used to show orientated movement remains the same in both diagrams.)

Figures 18–20 will show how this core orientating triquetra in Figure 3 becomes largely hidden as its dynamic twists to construct the analytical focus. This gives vastly different meanings to the classical terms mentioned in Figure 17. In each, a different sort of perceived reality seems external to the individual's felt identity. 'Reality' is projected out through the triquetral balance, tipping the inclination of thoughts and values towards self-interest in the 'personal identity' circle. Intra-action is lost; action on or towards the perceived reality is favoured instead. The *three dualist worldviews* that derive from I-it analyses are offered in the sequence of Materialist, Informationist, and Vitalist. The valuable analytical perspectives, which could be reintegrated into a triune whole if they had not been pushed to dualist extremes, are respectively called the Structuralist, Individualist, and Collectivist perspectives on life.

Shifts of worldview within the same brain can happen rapidly. Data is sorted informationally into different patterns among the neural networks. This illustrates how people can 'convert' from one set of firmly held beliefs about life to another that was previously incompatible. When life experiences are odd or strange or unnatural, and they cannot be integrated into former mental constructs about reality, shifting into one or both of the other two primary analytical perspectives may improve the capacity for exploration to find new creative order amid life's diversity. The triquetral concept of informenergy can sustain personal identity through the potential phase of turbulence that may be associated with transitioning from one organisational state to another.

Each analytical perspective alone can provide a view into the *whole* of life. However, each is incapable of giving a *full* description of how movement and diversity arises in life. Therefore, each alone cannot describe how healing into wholeness can be achieved. This is because the notion of an *individual self* with optional relatedness pervades all three perspectives equally. I-it mentality veils the personal intra-active wholeness of conversational orientation. For the same reason, the synergic informenergy nature of energy-matter understood systemically is veiled, concealed when individuals secure their own identities over against a perceived objectified essence to life. When a person accepts that these essences are only temporary ways to explore the relatedness that is already there changing in qualities at heart level, they may feel better able to relax into a healthy state of belonging. Life is a continuously emerging mystery shared with others. To contribute one's story, as seen from one's unique place in the extended conversation of life, will add great value to adaptable, intra-active life.

It is important to recognise that within each fully reduced worldview, a person can retain *internally logical consistency and rationality*. However, this internal consistency can only happen by wilfully choosing to exclude certain thoughts and factual observations from a perceived, believed 'reality'. That process of exclusion to retain internal consistency can be overcome, however, by converting the informenergy needed to sustain barriers into the movements that make fuzzy liminal zones of interfacing boundaries. It may feel risky to do so at first, but the investment of time to build relationships with those who see life differently will eventually build a more secure future for humanity.

Hypothesis revision: how conversational orientation is constructed

Three sensory association areas of the cerebral cortex construct patterned impressions of space, time, and experiential presence that pass forward and integrate *in the motor-planning cortex*. These impressions orientate active and stative responses within the wider ecologies of life. Each of the *primary constructs for orientation* is a dynamic synthesis of two of the three *primary sensory features*—form, change, and relatedness—extracted from all the distance senses for orientation:

- Space is synthesised from the *relatedness* of *patterned forms*.
- Time is synthesised from the *changing* of *patterned forms*, either cyclically or as progressive movement.
- Experiential presence as object recognition is synthesised from the *changing* qualities of *relatedness* that appear in consciousness.

In these association areas, the neurons in the surface grey matter (cortex) of the cerebral hemispheres are arranged as columnar tubes, perpendicular to the surface. These tubes are NOT the end point of higher thinking. They are stacks of processors like transistors, or tubular bells, that connect both horizontally to each other and down into the core limbic social brain, so that ripples of informenergy move through them to bundle their activity into organised movement. These neuroplastic bundles of activity then enter *feedback loops* that also connect widely across the cerebral cortex. Their feedback patterning also resonates with memories, embodying past interpretations into ideas that become associated with ongoing sensory phenomena.

The brain's higher analytical cortex receives *only sensory inputs that have been pre-selected and modulated*, however, through filters in the core social brain. These filters embody anticipation patterns for the probable sensory feedback expected to follow the planned motor activity. The analytical cortex is therefore primed to work only on whatever it already knows or anticipates. Therefore, self-interest feeds back its own self-affirmation. A wider interest in life, however, may feed back curiosity to magnify and explore that which does not fit with the expected.

Areas of psychological interest thus may focus attention involving the whole brain's activity patterning. The hypothesised proposal is that when people become interested in a feature of life, sensory features in that focus of attention are sent up through the centres of the cortical columns in the sensory association areas. Here they are modulated by the sensory inputs for orientation that are sent to Level III in the processor column walls for this purpose. Partially orientated sensory data emerges at the top from the centres of the columns in each of the three sensory association areas, which then pass (a) forward to the motor-planning areas for final integration around that person's conscious planning of their activity and rest, and (b) down by cortico-limbic feedback loops to prime the core brain's anticipation patterns.

The triquetral principles of projection

The three I-it modes of reduced personal identity are proposed to arise when a person's attention is drawn away from intra-active living (conversational orientation) to focus *on* the changes of life, or the qualities of relatedness in life, or the patterning of forms in life. In these situations, the relevant primary sensory feature for orientation would *also* be sent up through the centres of the sensory association area cortical columns as well as to the column walls. The modulating effect for orientation of an identical sensory feature in the column wall would then be cancelled out by that same feature within its centre. Only the other half of the orientating pair in the wall would have any influence to shape the context for subsequent planned action or rest. To understand the impact of this reduction on the way substance is perceived and valued, we shall need to look at the specifics of the three primary analytical perspectives.

The strength of this hypothesis is that all three reduced mindframes are indeed identifiable in the history and current dynamics of societies around the world. It is supported also by the fact that *conversions* of belief between these incompatibly different mindframes can happen without brain injury. They occur simply by repatterning connections within the same brain. Vastly different ways of seeing and valuing aspects of life can occur *overnight*, indeed can occur during 'A-ha!' moments of a thinking day. The reason this rapid repatterning can occur is that each worldview is *neurologically as justifiable as the other two*. Each uses for its *assumption base* simply a different pair of the three equally important features of sensory input needed for fully conversational orientation.

Conversational orientation (the me-you or I-thou mode of living) converts into the dualist I-it perspectives for analysis in this way:

- Each pair of primary sensory features for orientation becomes the assumption base that determines the inner 'I' or 'me' of "Who I am". This reduces personal conversational wholeness to an analytical *individual identity* mode.
- Each projected primary sensory feature for orientation that is sent up the cortical column centres in the association areas because it has become the focus of interest reduces to the *essence of a perceived objective world*—the external 'it' of "What it's all about?"

The logical deductions that follow from these different assumption bases generate different impressions of 'outside' and 'inside', and 'set aside' in each of these perspectives. This is because, to sustain integration within the reduced dualistic worldview, some features of life experience will need to be excluded from awareness, thought, will, or emotion. These worldviews cannot be empirically disproven within their own terms, however. This is because the proofs lie outside of reason as perceived from within the mindframe. However, neither can all three analytical worldviews be fully integrated with each other. Some deductions within a mindframe are wholly incompatible with those made within another from its own assumption base. Unity amid this diversity can only be found when people accept together, as mentioned before, that it is healthy to have a sense of mystery, and that awe and curiosity about *what we do not know* can be a uniting feature of life for humanity in the fuzzy boundaries of our liminal zones.

***

So now, following these *principles of projection*, we can start to compare and contrast the three analytical perspectives. To do so, first identify the basic triquetral shape in the following diagrams, Figures 18–20. It is easiest to start at the projected 'My Perceived Reality', where the triangular feature of interest has become the perceived essence of life. For example, in Figure 18 the red (mid-tone in greyscale) corner has become the separated red triangle representing an objectified form. A sample of the variety of ways this may be mentally 'seen' and named is shown within that triangle representing the externalised essence in this type of perceived reality. In physics, these would be perceived as particles. These objectified notions may be filled out by the terms mentioned in Tables 3.1 and 3.2.

The remaining two corners and central phase space of the triquetra are now within a circle that represents 'My Identity'. Its central zone now characterises that individual's core identity, but this is reduced to only the two remaining qualities of fully personal identity.

The Structuralist (Materialist) analytical perspective on life

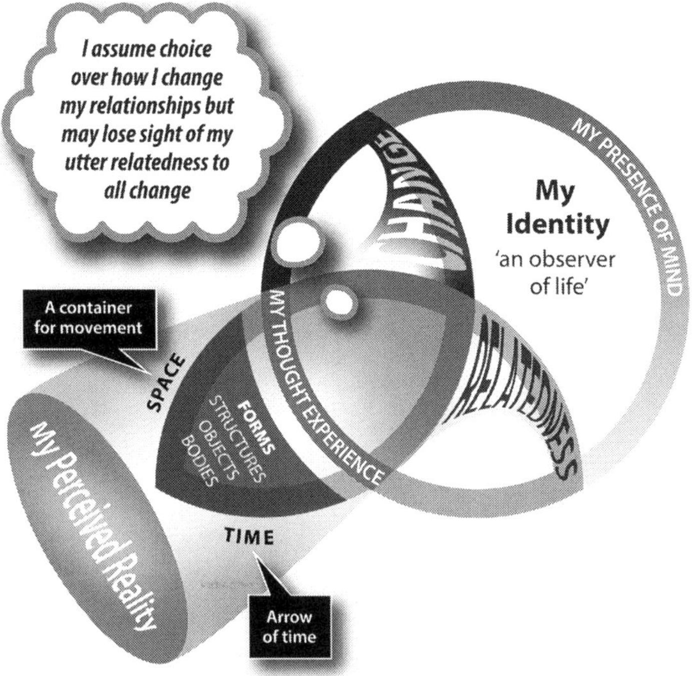

Figure 18: The Structuralist (Materialist) analytical perspective on life

To construct this analytic perspective, *form* sensory inputs are no longer diverted only to the cortical column walls to be paired along with change and relatedness, but now are *also* projected up through the centres of the cortical columns in all three sensory association areas. Here they mix with all the other interesting qualities of the sensed and memorised world. Form data now emerges from the tops of the columns as a feature of the external world modulated by the orientating inputs to Layer III of the column walls. In the temporal lobe sensory association area, form is fully modulated by paired changing relatedness qualities in the column walls, and so emerges as fully affirmed *solid objects* in the world. In the other two sensory association areas, however, the form in the centre of the columns negates the form that simultaneously has been projected to the orientating column walls, and so becomes modulated only by relatedness in the parietal lobe (form and change constructs *space*), and by change in the infero-medial frontal lobe (form and change constructs *time*). Therefore, emerging from the tops of these columns is the notion that space and time are also features of the external world.

In this analytical perspective, the world and the universe look like objects moving in space and time, which is the scientific materialist perspective on life.

In Figure 18, the analytical individual's mind in this perspective is a non-spatial Observer of life. Personal experience becomes a reality of thought, and the personal identity becomes a 'presence of mind'. The triquetral powers retained within this presence of mind are to change the qualities of my relatedness with the physically separated objective world. Thought experiments become as realistic as physical experiments, but the unique feature of the character who exercises these powers of choice is a moral separation of responsibility for the impact of personal choices on the objectively physical world. This happens because there is no concept of the utter relatedness of all changes among which this I-it mode of personal identity is intra-active. There is separateness instead.

Here we see the origins of the Enlightenment view of the universe, as solid objects moving in a space-time container with momentum, and interacting by bouncing off each other with conservation of their energy. This perceived reality frames the empirical planning of action. This mental frame for frontal lobe planning therefore embeds values in the core brain that are associated with that analytical worldview. The triquetral icon, now unbalanced in this way, is tipped by these filtering values into an inclination for thought arising from observations of sensory feedback from the separated physical and social ecologies. *Form* data of interest in that sensory feedback will therefore continue to be magnified and to pass on through to the *centres* of the sensory association area cortical columns, and on from there to the frontal lobes. A cyclical trap now closes the system.

The personal wholeness system can reopen easily, however, into an open system at any time by conversational connection with people who see life differently, and by activating a healthy curiosity to reassess former ideas and values. This process of letting go of former values would be helped by also activating an Emotional Logic understanding of unpleasant loss emotions. This could prevent whirlpools of loss emotions from trapping people with regrets.

The thinking individual in this view becomes an Observer of life. In this dissociated state, thinking individuals are easily isolated from others, so they may struggle to define 'who I am' and find themselves filled with tormenting emotions, or a gathering darkness of soul, seeking medical, psychological, or spiritual help to restore connection. A culture of therapy can grow around this materialism, rather than a culture of personal growth through heart-level relatedness and mutual synergic responsiveness. This is evidenced in the growing tide of common mental illnesses, relationship instabilities, and socially disruptive behaviour that can accompany a materialistic approach to society.

This analytical perspective creates a dualistic mode of personal identity that can be identified in the teachings about substance of Democritus in fifth-century BCE and of Epicurus in second-century BCE Greece, who built a moral, or rather an amoral, philosophy on the notion of atomism. Not shown in Figure 18 is that the classical terms 'soul' and 'spirit' have disappeared. They have become meaningless notions in the presence of this analytically reduced empirical mindset. The power of choice of a Materialist is found in Observer mode only. But we must not forget that individuals can and do create their own mixtures of the three primary analytical perspectives to construct their unique worldviews. Pure Materialism is rare, and can easily moderate into the Structuralist view that values order and predictability but does not need to impose it on everyone. More flexibility in a fully triquetral mindset can be easily restored by first recognising that *personal values* have a physical substrate of embodiment in the inner ecology. Then connecting personal values with mutual responsiveness in the social and physical ecologies can restore a sense of belonging as a fully responsible and influential person with others, sharing environments and adapting together to their changes.

The arrow of time and objective time

A remarkable consequence for understanding *time* follows from these shifts into the I-it modes. Three types of time result, not the classical two of subjective and objective time.

In the Structuralist (Materialist) worldview adopted by an Observer of life, time has its own reality as an externalised context against which the movements of solid objects can be compared and measured. Time in this experiential perspective observes differences between past and future states, which seems logical *within* this perspective, but which ignores the implications of the fact that the mathematics of related physics equations works equally well in both directions of time. Within the Materialist perspective, however, theories of movement focus exclusively on linear cause-effects observations of objects.

This *linear time* sequence of cause and effect may be pictured as the arrow of time, from past to future. Because these changes are comparative and measurable, the notion of 'objective time' has arisen, which is differentiated from 'subjective time' within the Observer status. However, we shall go on to look at a second analytical perspective, the Individualist (Informationist) perspective on life, in which time is also projected out as a context, for the changes of life in that perspective. Time in that analytical perspective looks like a timeframe within which a certain measurable number of changes need to occur to tell a successful story, with a beginning, middle, and end. Both linear time, and timeframes, are comparable and measurable, and therefore both contribute to the notion of so-called 'objective time'. However, in triquetral cosmology and psychology they are significantly different. Belief in a notion of objective time is a value statement only held within these two analytical perspectives. In the New Science view of quantum gravity, that notion of time as a context for the movements of objects or changes has been abandoned, more of which later. Belief in objective time requires one of those 'set aside' states of *chosen disbelief* that are required to live in an analytical perspective. Any measurement requires agreement between subjective people to limit the focus of their attention to the agreed way, otherwise measurements are meaningless. Events and understanding can happen in people's subjective experience outside linear time sequences; and timeframes have fuzzy boundaries that are bridged by anticipations and memories.

Objective time is only measurable by making comparisons. Subjective time involves making personal choices that alter social processes. To recover their underlying triquetral synergy as we progress to a New Science view of movement, objective and subjective time can be creatively merged into intra-action by focusing instead on the qualities of relatedness that are needed to make life-enhancing choices.

We return now from letting go of the objective-subjective time distinctions to considering the impact of narrowing to an 'arrow of time' view in the Materialist perspective. It is important to recognise that space and time have both become conceptually objectified in this perspective and attached to a form that has been rendered *static*, internal to itself, by the objectifying mental process. An external causative force is therefore needed to alter the object's momentum, or to split the object into its parts or particles, or to change its spin. These movements all imagine the particle to be internally static in this atomist mindframe. The causative force therefore acts on a static image of a form *within a wider picture of patterning change*. Effects follow consequentially from causes in this linear time sequence. A 'set aside' conundrum that results is the unanswerable question of what force set 'it' all in motion. The triquetral cosmology that will be presented in Part IV gets inside forces and objects to offer its own solution.

To such a static noun type of imagination, it is important to calculate how a force acting on an object can bring about an alteration of momentum and spin. Sir Isaac Newton developed a type of mathematics called calculus that describes an infinitesimally short period of time called a moment in time. This moment in time itself becomes a static noun called a present moment in this reduced worldview. This is followed by another, but differently patterned moment, causally connected in some way to the previous one. In this analysis, objective time looks like a string of slightly different pictures in frames, extending from past to future, with the present moment constantly in the middle but progressing along the line, rather like the old cinema film reels. In the string theory or M-theory of quantum mechanics, each momentary frame is called a 'brane' progressing through a bulk of a number of 'hyperdimensional spaces', which suggests that M-theory sits within the Structuralist perspective on life.

This 'arrow of time' view is thus a foundational assumption of Enlightenment Science to allow measurement. The mathematics of physics equations works equally well in both change directions, however, so the search has been on among materialists to find the particles that make time move in only one direction, as their theory would predict. If time is seen as a long string of present moments, the concept of eternity simply becomes a never-ending string, which many people subjectively find quite scary. The logical thing to do then is to set aside thinking about such a distressing mismatch.

Enlightenment Science makes no allowance for 'final causes' in the features of life that it measures in linear time. The idea that future moments might have a teleological purpose that reflects back into shaping the present moment's decisions is ruled out by the Enlightenment scientific method. The New Science has confounded that view on a cosmological scale. The Enlightenment view is still of great value in making sense of our lived localities, provided that the *choice* to focus on a locality is always remembered when interpreting how generalised the measured results can be. Observations in one locality should not be used to limit people everywhere. Final purposes, as for example in building a sailing boat that will not sink, need a storytelling *timeframe* as imagined by an Individualist Creator of life to explain how final-cause design could also shape the present moment through the other three Aristotelian causes: 'material, formal, and efficient causes'.[2] All are needed to fashion the parts of the boat into a whole, using 'scientific principles' in technology, and psychospiritual principles in the teleology of sailing. Final causes *are* logically consistent with the Individualist (Informationist) *timeframe* and the Collectivist's (Vitalist's) *eternal now* worldviews, but in slightly different ways to be explained later.

Letting go of a narrow and exclusive 'arrow of time' view of life may generate much grief for people whose whole life education has been shaped by Enlightenment Science. It may deeply challenge value systems that are currently dehumanising life, to everyone's benefit eventually. Therefore, understanding the other two primary analytical perspectives, the need for awe and mystery to hold all three together, and the Emotional Logic method to support healthy adjustment, will mark out the paths to explore ways forward.

Psychology and cosmology in a Structuralist-Materialist view

Mismatches from the anticipated order of this observed world will seem odd. For example, dreams that foretell the future, and the telepathic remote transfer of *emotionally intense knowledge* at an intuitive heart level from one person to another, will probably be excluded by choice from the analytically observing mind in Materialist mode. If people habitually think in the Materialist frame of mind, phenomena such as these will be difficult to fit rationally into a consistent mental order. These are uncontrollable and non-replicable features of life that would normally be discounted as random material interference, or subjective imaginations, or worse, as intentionally manipulative deceptions to confuse others.

There is another clue that there is something wrong or incomplete in this reductionist, dualistic view of the physical universe, however. If the physical universe is ontologically real and logically linear as cause-effect in itself, then the objective form of time that is the basis of each present moment should be really real, and knowably so by a momentum effect in a real brain. But the human experience is that the present moment is fleeting, like a vapour dispersing and forming and transforming in its transience. We cannot grasp the present moment in the same way that we can grasp a lump of clay, unless grasping that clay *mindfully* by a wilful choice in some way parallels grasping the present moment triquetrally to live in it. We can commit *by choice* to behaviour within such a present moment, however. Those choices are framed by the values that we construct moment by moment by feedback in the inner human heart's sensory filters. There is subjectivity behind the chosen values even of modernist science. The New postmodern Science of systemic emergence acknowledges how relational intersubjectivity is integral to the structure of the cosmos. This places personal choice at the heart of lived experience. Personal choice also determines the *comparative measurements* that both modernist (Enlightenment) and postmodernist New Science researchers take as they explore objectivity within their vastly different concepts of the cosmos.

The New Science discovery that the universe is *shaped from within* by loop quantum gravity, not shaped from the outside by a space-time container, has pushed into centre stage a dilemma about the need to extend our worldviews out from our localities to a cosmos that births life from within. Human beings have once again become intra-active gardeners and caretakers and midwives of life within a limitless, open-system cosmos. Human beings are once again morally responsible choice agents. We are mindfully embodied within the movements of the cosmos. Like actors in the ancient Greek tragedies who wore the masks of 'persona', human beings play their parts within the unfolding story, but without a pre-written script. Each person writes the script as we write our signatures across the world, and now across the solar system, and exploring beyond. This new fact creates a problem for those who like to shift responsibility from themselves for the quality of their shared environment. By remaining in the Materialist mindset habitually, individuals can lose sight of their heart-level connectedness to shape life by extending their sensitivities. By limiting attention instead to only the efficiency of 'natural processes', and the pleasures of an observing, irresponsible mind, the ethics of decision-making in a Materialist mindframe has become situational, local, and individually pleasure-based. In triquetral terms, this is a recipe for instability.

The internal logic of the Materialist perspective, seeing a clockwork universe, has generated the most remarkable improvements in *physical* quality of life across the globe. The atomist idea of internally immutable particles *does* identify how much of life *can* behave predictably in controlled environments and in engineering. This dispassionate view of matter moved by an externally regulating mind has worked to a significant extent for the common good—albeit with huge continuing inequalities of access to food and energy, huge problems of pollution, widespread access to destructive weapons, and an increase in violence, mental distress, drug misuse, and illness through relational disruption and isolation. To the logic of Materialistic dualism, we can resoundingly say, "Yes-and-no-no!" We shall need to move on to look at the equivalent benefits and problems of the two other primary analytical perspectives before a balanced solution can appear.

***

To round off this description of the Materialist mode of I-it reduced personal identity, certain important concepts can be summarised and marked out as different from the way they seem in the two other primary analytical modes. These differences are all subjects for discussion.

A summary of the Structuralist (Materialist) analytical perspective
In the Structuralist (Materialist) perspective on life, an Observer of life will see that:

- **Reality** *is* matter.
- **The dualist essence of life** is *energy-matter understood as particles.*
- **Space** seems to be a container in which matter has its own momentum.
- **Time** seems to be a linear process (the arrow of time) as an inevitable causal sequence that shapes material processes from past to future only.
- **The dualist personal identity** is an Observer *of* life, the *presence of mind* seeming to be a different essence from

energy-matter, but with no explanation for how these two essences interact.

- '**Soul**' is an incomprehensible term, for which 'mind' seems a fair substitute.
- **Spirit** likewise is a mysterious and unmeasurable concept, best considered an archaic mode of primitive thinking that has been superseded by our deep knowledge of the physical structure of matter and energy.
- **Body** is self-organising matter under the influence of DNA and RNA, which has a mysterious connection to mind, and is of less value than mind. Transplants of body parts from person to person are considered reasonable to sustain mental life, but when mental life ends it is reasonable to end the body.

The final point to be making, which we shall return to in Chapter 12, is that the value systems accompanying this mode of thinking, and the other two equally reductionist modes, have a profound impact on ethical behaviour, and on the shape and health of human society as populations and individuals adapt to ongoing environmental change.

The Individualist (Informationist) analytical perspective on life

In Figure 19 we see a different analytical orientation. When a person takes a mental step back from conversational orientation, having become fascinated by the *changes* that are ongoing, the concepts in the blue (dark) corner of the triquetra become separated away, creating a different type of perceived external reality.

Change data are now being directed up through the centres of the association area cortical columns and on to the frontal lobe motor-planning areas, with a different projected context for life. In Figure 19, change has become objectified as an 'it' of chaos or of organised processes and flows that are constantly and perhaps unpredictably influenced by people's choices. All of this turnover occurs within a projected context of real time, which is experienced as potentially stressful timeframes for the completion of tasks that 'I' have to do. Objectified changes are now noted everywhere in timeframes, as processes that need organising to avoid potential chaos. Life becomes busy. It comes as a surprise that life goes on even after a deadline has been missed. Unpredictable people are making unexpected choices that may disrupt the smooth flow of ordered life in any setting, all simultaneously, which adds to the chaos. They need engaging into organised patterns of activity to make things happen in time, before the deadline. In physics this would be called random activity, or interference, potentially disrupting laminar flow during an experiment.

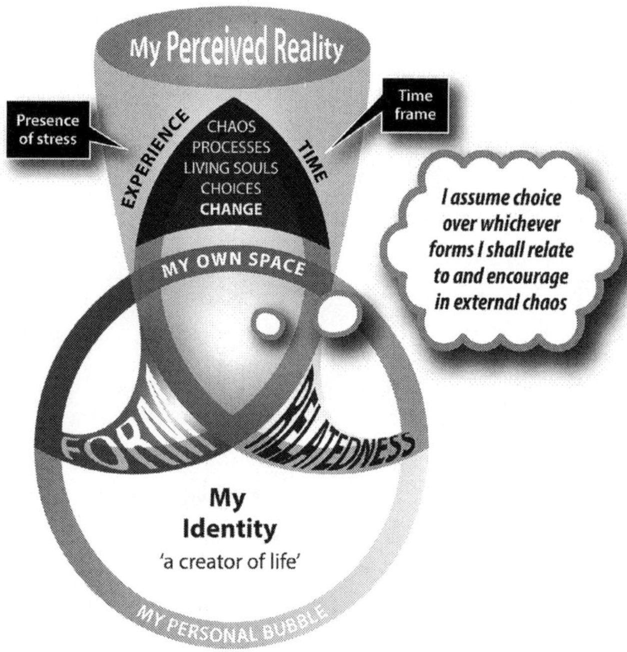

Figure 19: The Individualist (Informationist) analytical perspective on life

This externalised chaos is now set over against a circle that represents 'My Identity'. Inside the circle, the central zone of the triquetra that orientates that individual's core identity is now integrating only form and relatedness data effectively in the cortical column walls of the sensory association areas. The two other orientating constructs, as described previously when considering the Materialist analytical perspective, have their change elements negated by the content of the columns, so that both *experiential presence* and *time* get sucked out into 'the real world'. The clock thus gains authority to determine this person's choices of behaviour and values within their inner world of individual identity. The form and relatedness features that now characterise the inner world looking out on potential chaos create a notion of inner space, my personal bubble, within which 'I' have the authority to shape life as I want. "I have the power of choice to allow people in, and to exclude other people, from my personal bubble. I have the power of choice to make life as I want it to be shaped, and nobody else has the right to tell me otherwise." The notion of the person as a social creature has been reduced in favour of individualism and information generation. The notion of 'personal' no longer means my unique contribution to the public, but now means private and not to be questioned or intruded upon.

The triquetral icon is now unbalanced. This analysing individual will feel it is their role and right to bring order to, or to sustain their perceived order within, the perceived external potential for chaos. The relational quality between inner self and external perceived reality becomes one where there needs to be a transfer of information (noun) to alter the patterning of life 'out there'. Information becomes objectified to bridge the perceived gap between outer world and inner self, creating a dualism in which an information generator is remote from an information receiver, and in which the transfer of objective information is subject to random error. It, the information, traverses the gap to seed order in the midst of the randomising 'pull' of entropy. Information (noun) has become the essence of life.

Personal space becomes a mental 'workspace' or workbench for creative mental activity, which is organised and pushed out into the world within a self-constructed timeframe in which its information is imagined to transform the processes ongoing there. Personal presence does not need to play any part in this perspective on life. It is the essence of information that does the job, which leaves 'me' as a remote '*Creator of life*'. Within my personal bubble, my mind seems to have the power of choice over whichever forms of life or ideas I choose to pull together, to name as I choose, to magnify or remove, to imagine, or reorganise, to create a coherent or interestingly patterned order that says something of my influence in the world and in society. This may feel like 'making sense of the world', or 'finding meaning in life'. However, an opposite dynamic can also follow, such that some people choose to hide within their personal space because the chaos seems overwhelming or hurtfully dangerous, choosing to create their own inner safety in their hideaway there.

Either way, the individual believes that they can best preserve life by generating their own changes. It is not inconceivable that this frame of mind could have generated the classical notion of a self-existent soul, separable from the known world. People moved by information in the external world could have their own separable souls, which make people seem distinct from each other and responsible entirely for their own state of being in the world. This is the opposite notion to the triquetral view of soul, in which one person is fulfilled as a living soul when in synergistic relatedness with others, materially and spiritually. Likewise, the classical notion of a relational spirit is now turned inwards, because objective information is turned outwards instead. Individuals may be called spirited, or spiritually strong, if their inner energy moves them to actively get on and create something that influences (or controls) the outer world's chaos. In classical terms transferred from Figure 17, the personal bubble is reduced so that only spirit and body are contained there (the relatedness of patterned forms). In this resulting inner ecology, an Individualist Creator of life 'embodies' their generated information as their *idea,* fashioned on their mental workbench before transmitting 'it'.

What does this look like in our present-day digital world of Information Technology (IT)? It is a pressing question that needs a practical solution. Digitalised information has replaced analogue personal synergic relatedness. But even more than that, in the face of rapidly advancing Artificial Intelligence (AI) and robotics, people will need to explore how they can stand together effectively as stable domains, in mutually supportive and adaptive groups in which to become fulfilled in their lives, rather than being dominated by robotic efficiency. This Individualist (Informationist) analytical perspective sets aside the need to be present for each other. Remote communications are good enough in the present timeframe.

To answer this question of how this looks in our present and next three decades, it will help to recall that the transfer of information (noun) needs some sort of carrier wave that is modulated at source, and is responsively received at a remote location. In the Individualist dualist view, the information pattern itself is the essence of life. *It* could pattern *any* carrier substrate. The nature of the substrate carrier of information is insignificant in an Individualist way of thinking. Any 'medium' will do for the transmission of objectified information, provided that it achieves the individual's chosen purpose. Notepaper messages, beautiful objects as gifts, online protocols, radio transmissions of music, optic fibre modulated light carrying conversations across the oceans in cables, even the esoteric transmission of information from mystical sources, all will do, provided an intended change follows somewhere else. Avatars and screen names are adequate to embody the individual's information across The Cloud. Activating a digital robot could be satisfying for an Individualist. For a triquetral thinker, however, analogue pleasure might, for example, come from guiding and conducting a choir to sing harmoniously.

In a balanced triquetral mode of thinking, the nature of spirit *is* the synergic, even synchronistic, quality of informational relatedness *between* source and receiver—the carrier wave, so to speak (although not ontologically so, as is the error of Vitalism). Triquetral synergy emerges as a resonant harmonic into an inner awareness of movement that is between one and another. Information (noun) is not carried *on* spirit in a triquetral mode of thinking; and neither is spirit an essence that carries information (which is the Vitalist mode of dualism). In a triquetral mode of thinking, the mutuality of movement across space and time is *informational spirit,* connectedness. However, that relational primacy is devalued by an Informationist. The Informationist is mostly concerned with the outward transmission of their information, which they have fashioned and created. Their concern is not in how it is received elsewhere. That is 'up to them!' The notion of relatedness, triquetral spirit, is absorbed internally by an Individualist into their construction of *patterns of ideas to transmit*. Some Individualists are more spirited in this way than others, but may then drive others into stressed states with too much information without realising the impact they are having. Sensitivity to feedback may be lacking when the construction of interest is within. All of this affects the values that are embedded in the core brain's anticipatory filters, creating an attractor state that may be difficult to move on from, towards a more systemic view of balancing curiosity in life.

Timeframes

Information Theory drives individuals to measure effectiveness objectively, which means using linear time sequences to compare changes of state. It can lead to a close cooperation between Individualist and Materialist worldviews focused on the notion of objective time. The Individualist component of objectivity is to measure the length of the timeframe and the number of changes within it to bring about the desired change of state, although the sequence of these changes is in itself, in this view, unimportant. Objective management of people is interested in both these aspects, comparing physical and social systems with the clock to regulate them.

In triquetral conversational orientation, timeframes have an additional quality. They enable *relationship* to be known, which is a feature of subjectivity. Unexpected changes can emerge from movements that are relationally aligned. Sensitivity to the resulting changes can lead to a response that could bring to mind the synergic mutuality behind indeterminacy in relationship. The reality of life becomes less about information concerning the starting and concluding states of a process, and more about the often startling stories during the journey. Behind every behaviour there is a story, making every moment one of potential richness if curious. For humanity to survive, value should be attributed to the stories of rest and peace in stable domain groups, as much as to action and who dominates whom for a time, until equality and justice is restored amid diversity. Both types of story are needed for wholeness. In this, human beings are not weaker than robots; they are just human and different from robots. Human storytelling and story listening in small groups and large is more than sharing information, or instructing a robot. When people focus their values on efficiency, a synergic element is missing. That is why setting objective targets in timeframes in organisations is dehumanising, while recruiting staff into vision, with goals that are achieved by sharing the contributions of diverse personal values in the team, is more sustainable, adaptive, and organic.

Nevertheless, the Individualist analytical perspective adopted by a Creator of Life has had tremendous benefits for society, especially when its insights and innovative ideas and skilful artistry can be heard and tested and integrated into adaptive social wisdom. Creative arts add beauty and stimulation to an expressive life, which might otherwise become humdrum and overregulated. Intellectual Property can be defined with objective boundaries, so that the spreading effects of new ideas can be traced in timeframes, leading to reflective practice that empowers learning and restructuring. Improving productivity or profitability can go hand in hand with humanity in a learning organisation that receives feedback about human values for sustainability and growth. When balancing these many benefits for humanity with those that have unfortunate and dehumanising impacts through the digital data explosion and remoteness of life, there must be a yes-and-no-no judgement on becoming stuck in this analytical view, where information has become the essence as a noun, rather than a socially dynamic verb. The risks of unbalanced Informationism in the coming decades of environmental change are great, given that information (noun) does not directly move people's inner hearts. It is filtered by pre-set values. To become unstuck and moving, Individualists could choose to explore their feedback and reflective learning opportunities with those who differ and could actively seek the informational benefits of diversity and storytelling, exploring other people's personal values in the resonant unity of relatedness.

A summary of the Individualist (Informationist) analytical perspective

To round up this description of the Individualist mode of I-it identity, a summary follows of the important core valued concepts. Please remember that I am defining here the equivalent of only one of an artist's three prime colours, from which an infinitely diverse colour palette could be mixed in people's life experiences. Having defined all three, we can look later at their mixture in a conversational process with people who see life differently.

- **Reality** is the processing *of* matter as information.
- **The dualistic essence of life** is *information* packaged in a way that is believed can influence the repatterning of life.
- **Space** is my personal mental bubble interacting with other people's personal bubbles anywhere.
- **Time** is a stressing timeframe within which changes must be made to create or fulfil life.
- **The dualistic personal identity** is a Creator *of* life, bringing order out of chaos by releasing information that is generated from their *personal space,* either aggressively, assertively, or quietly.
- **Soul** is the activities and changes within a personal bubble that generate information and knowledge. Some people imagine that this process continues uninterrupted beyond death of the body, as the survival of an enduring soul in a different informational context from that of the material world.
- **Spirit** is informational generation within and transmission from the personal bubble.
- **Body** is the shaping into a recognisable form or pattern of any substrate that carries the information. It can move any and every level of life, from physical matter to virtual avatar, and to the functional impact of leadership of groups on the shape of society, which is a patterned body of people.

The Collectivist (Vitalist) analytical perspective on life

The equally valid third dualist perspective of this triquetral hypothesis generates a very different worldview and I-it mode of personal identity.

For a Materialist, *presence of mind* and individual experience is internal, while space and time are externalised.

For an Individualist, a personal bubble of *inner space* is internalised, while a real presence of timeframes is externalised, which stresses the individual's life and informationally stresses others rather than synergises with them.

For a Collectivist, personal *time* is internalised. The unique quality of life of an individual in this mindframe seems to be his or her emergence from eternity and return to it after having 'My Time' in this world. While space is externalised and objectified as real, it is pervaded with the real presence of powers experienced as moving through the world and the self and all humanity collectively. These powers seem to move life as a whole. They have the authority to shape the lives of all the individuals dwelling together within it. This is the Vitalist optional further development of collective thinking.

In Figure 20, conversational orientation becomes disrupted when a person takes a mental step backwards to take note thoughtfully of the qualities of relatedness that seem to move life between people and with the physical world. Given the informational shifting that happens in the cortical columns, the notion of relatedness could become projected out, as if a real feature of an ontological external world. Simultaneously, the concept of space is externalised as a real environment that is pervaded by a richness of relationality or connection.

In a conversational orientation, relatedness is knowable in a context of time or timing. When relatedness is projected out as the interesting feature of an externalised reality, the watchful self would internalise time as its mindful character in order to make sense of the watched features of relatedness qualities. The self becomes 'My time as a Participator in this life of relatedness'. The participatory feature of *my time* is living in the now, while being aware of all time in which now is the focus of lived experience. Time becomes the eternal now, which is a frame of mind very different from the linear view of eternity that a Materialist imagines, and from the optional timeframes for action that an Individualist will create a story within. 'My time' is emergent from an eternity of wholeness, and perhaps my soul or my unique life will endure through endless moments to return into eternity with a continuing sense of my participation in the wholeness of life.

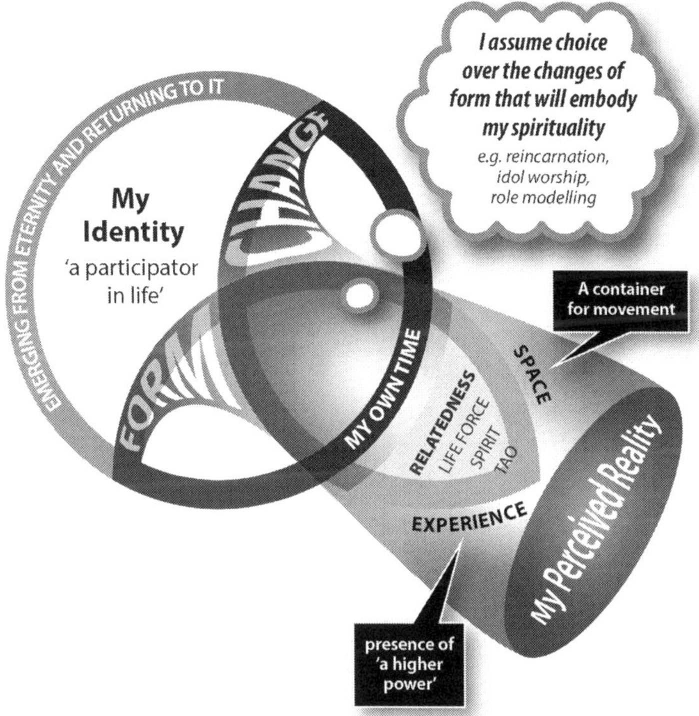

Figure 20: The Collectivist (Vitalist) analytical perspective on life

In the eternal now, an individual's appearance in life among others is like a traveller arriving, a temporary and vitally important phenomenon to be enjoyed with hospitality. Things and people 'come into their own time' to be experienced and lived. They emerge and they disperse, and their value for personal development should not be missed in that opportunity. This is sometimes called 'African time'. The *ubuntu* of communal respect and good manners places relatedness as a far higher priority in civil society in this analytical perspective than the pressure of prioritising timeframes for individual fulfilment, or for greater efficiency in linear timing. Final causes (purpose, teleology, fate) seem far more important in this perspective than efficient or material causes. Within this completely open timeframe of eternity, valid evidence of connectedness is considered in some circles to be prophecy and its fulfilment. The events are not considered to be boundaried purely by material time-sequenced cause and effect, or by individualistic self-centred choices. There is a higher agenda that crosses time and space to emerge into substantial reality 'in its time'. The same has been found by quantum physicists in the smallest scale of the material world where we all participate.

This notion of personal time is not the same as subjective time. Subjective time is the 'set aside' concept created by those who value the measurability of time or timing. The fully triquetral concept is called 'personal time'. It includes living mindfully in the eternal now, and adds intentional purpose that is best shared equally with others. Personal time is not about subjective feelings, as in that part of Individualism that revels in its unique subjective experience. It is the context in which *personal values* influence and are influenced by physical environments and shared spaces. Personal time is about social responsiveness, not pure individuality. Contained within the eternal now are the harmonics potentially of all past and future connection. Sensitivities to this, and breadth of vision, vary between individual people. But these features of character may develop if, in groups, families, and collectives, people choose to focus on naming their personal values to each other as they journey while telling their stories.

This enduring continuity of the self may become a quasi-ontological notion of an enduring soul or person, however. Figure 20 shows how the personal identity I-it mode of a Participator in life, who has projected out relatedness as a greater reality than their own unique experience, retains within its personal empowerment circle the characteristics of *change* and *form*. So a Participator in life, who is analysing how to thrive in the present moment in terms of qualities of relatedness, needs to consider how to transform their behaviour in the collective setting to better engage with that ontological relatedness that they so value. This Collective perspective is therefore the source of a *personal development* frame of mind. The Structuralist and Individualist perspectives do not in themselves inspire this sort of personal development, but rather invite people to seek their own empowerment over the world, or satisfaction in the way they shape it.

Eternity and reincarnation in the eternal now

Eternity for a Participator in life is not the linear sequential view of eternity that so troubles a Materialist. It is instead a relational state of wholeness. This wholeness must nevertheless exclude poorly developed souls to lower levels of 'waiting' and processing for development to 'higher' levels. Clearly, different cultures around the world and throughout history have conceived this state of perfecting the soul within their own diverse religious or philosophical languages. There is one uniting feature, however, that is important to recognise.

In this analytically reduced view, the retained powers of form and change by an individual add a sense of personal responsibility for growth of character. In this analytical perspective, therefore, returning from a lower level of eternity in a changed form for another round of perfecting character is rationally logical, given the assumption base of this analytical perspective. Reincarnation of the soul into a different body form is a justifiable deduction in this perspective. However, it is logically incompatible with adopting a Materialist perspective on the self and world, and so cannot in itself provide an integrating basis to discover the fullness and wholeness that is truly sought. For an Individualist, reincarnation can be an optional and situational belief, fleetingly depending upon the timeframe that holds their interest in which to count what matters before moving on. Nevertheless, the creativity and originality inspired by individualism is a triquetrally vital contribution to the overall adaptability that people in groups ultimately need in a wholeness of unity that intrinsically generates life-giving diversity in its localities.

A cultural belief in reincarnation is not the only way that this Collectivist perspective on life can express itself. The Christian notion of resurrection makes this more of a one-way process that leaps forward into that wholeness. Ancestor worship and the notion of spirit guides are other expressions of that collective perspective, which includes informational synergies between different states of life. In the absence of a spiritualised context, secularists experience this collective way of thinking as a desire to leave a legacy for subsequent generations, leaving the world a better place than it was on their arrival. Other expressions of collective spirit are idol worship inspiring an inner transformation into a desired state, and role-modelling on people highly respected. These are all expressions of the same conceptual mindframe, which can be mixed in an infinite variety of ways with the aspirations that follow seeing life from the two other analytical perspectives. The result could be more flexible patterns of relating with a curiosity that explores how Structuralist, Individualist, and Collectivist thinking can all contribute equally to adaptive living as our shared environments change.

Vitalism

A Participator in life analysing their experience in a Collectivist perspective may go a step further and open the door to objectifying the concept of relatedness rather than seeing 'it' as synergic movement in a liminal zone where different dynamic systems meet. When objectified, the notion of a vital force can captivate the mind and embed itself in personal values that affect communications and behaviour. Throughout history this tendency has emerged in varied cultural ways, for example as Chi, Prana, Platonic Ideals, Hegelian spirit, and more recently as esoteric energy or energies. If this connectedness is seen as the prime mover of life, beyond the individual and coursing through the individual, this essence or substance may be imagined to be pre-material and eternal, moving into 'my time' even before the appearance of matter. A uniting feature of all these culturally diverse terms for the same concept is that they all course through space; and they all have a quality of being a 'higher power' over life and the individuals who dwell in 'its' presence for a time. This vital force is imagined to pass through all bodies and minds and hearts pre-materially, with an authority that commands conformity for healthy life to flourish. The individual's inner heart is usually perceived to be at war with the purity of the life force, and in need of healing from inner corruption.

In conversational orientation, energy-matter is a *synchronous* emergence of substance and understanding. Relational synergy is described as a unitary concept of informenergy that meaningfully connects diversifying localities in wholeness. Life is the integrating movement behind all that seems static. These notions are significantly different from the notion of an esoteric, ontological essence that is itself eternally unchanging. In conversational orientation, movement allows all three analytical perspectives to emerge and then to re-blend in ways that enrich collective adaptability. That multi-analysis and re-blending is the way that good character develops, which then feeds back and contributes to the life of the whole in eternity.

In the Vitalist perception of reality, solid matter is only an interference pattern, like an aura that develops a density akin to material, but is not existent in itself. The values or inclinations of the inner heart thus shape the personal soul, which then interferes with the absolute life force that is coursing through it. This creates not only the illusion of matter, but also for some people the notion of yogic progression of the self beyond matter, perhaps to the formation of a spiritual body.

An alternative collective view of souls is the metaphorical description of peoples or nations as the oceans (or waters) of humanity. This is a strongly Collectivist view of relatedness in society. It can be misused to apply pressure to conform to behavioural norms, as seen in caste systems. But conformity is not the same as unity of heart. Unity requires diversity for there to be life. The important feature of collectivism to balance is the mutual *equal valuing of differences* for synergy to emerge, so that informenergy movement can reform life.

Integrating the ego by synergic resonance with others

A curious feature of reincarnation thinking is that it tends to be time-sequenced towards progressive maturation, as if there will be a moral advancement of society. People may talk about recovering memories of previous lives, but we hear less about looking 'forward in time' to future lives. The triquetral concept of synergic relatedness across all time, space, and substance is that the core communicational status of wholeness is vibrational movement between localities of that implicate, pre-orientated order. Personal values are embedded here by the meeting of feedback cycles of the inner ecology of an embodied person with the wider physical and social ecologies.

If this picture of embedded personal values is true, they have a vibrational impact in the implicate order prior to any emergence into space-time-substantial explicate order. In which case, harmonic resonance between persons with harmonising values could occur in the implicate order, and emerge into explicate human experience in a number of places, times, and physical or relational settings when the mind is open to movements of the inner heart.

Participation in life is not passivity

A consequence of belief in the individual soul's dependence upon an impersonal life force for spiritual progress is that the valued focus of attention narrows onto one's own karma, which may be viewed as the sum of a person's actions in this and previous states of existence deciding their fate in future existences. Karma accumulation is therefore the linear equivalent in the Collectivist spiritual perspective to cause-effect process in Materialism. In a triquetral view, conversational orientation would consider karma more as an evolution of personal values from one systemic state into a more life-enhancing relational state that benefits the dynamics of all who constitute the social system of implicate wholeness.

Taking a too individualistic view of the soul's progress with karma may result in people looking on another's suffering and choosing not act to relieve it with relational mercy. Participation in this view has limits, boundaried to the individual's soul, even when environmental turmoil such as earthquakes impacts life for others equally. There is an equally Collective but non-Vitalist participatory belief that allows merciful action to be a path to ultimate tranquillity in the implicate order. It requires a further transition from analysis into a triquetrally lived mindfulness that balances dispassionate materialism to relieve suffering with a belief in individual creativity in action through elevating qualities of relatedness to a sacred place. This balance moves beyond passivity and self-centredness into the mindful choice for action and communication to transform environments shared with others. In this way, karma may come to rest in the wider implicate whole.

The Collectivist perspective, therefore, does not inevitably drive people towards a Vitalist philosophical or spiritual belief in either karma or reincarnation. Its impact on qualities of life will vary depending upon the physical environment that occupies the interests of the enquiring mind. In family, workplace, and culturally artistic settings, for example, socially acceptable relational rules or guidance will apply in one setting that nearly everyone intuitively knows will not apply in another. We can slip from one set of relational qualities to another depending on a sense of geographical location, although there is also that underlying sense of one's core personal identity or character that transfers between them all. Group identities can emerge that endure despite a people being moved by traumatic events into a diaspora. Activity in remote or extended communication patterns then keeps living systems adapting and evolving as environments change.

When love in this collective worldview turns from its joyful state of mutuality into its grieving mode on facing loss, unfortunately both secular and spiritual interpretations of this analysis of life's relatedness can start to be driven by fear and feelings of insecurity. Mythical spirits can be elevated to the status of gods, with externalised powers that seem to influence or even control human lives. This unfortunately leaves many people open to paranoid forms of worship and sacrifice to appease the unknowable. Life can become full of dread, rather than filled to overflowing with cooperative love. Small groups may break away to fulfil the collective feature of their natures, but only by excluding others who are less well known. The potential for charlatans to play on people's fears can then break the *creative collective* nature of this view of life. However, the strength of this Collectivist mental frame that gives it its clear yes-and-no-no status is that something of value is recognised in the heart of human nature to which the rational mind can aspire in its choices. That something endures through time, even beyond the timeframe of any individual's life. We are all personal substates in something bigger.

A summary of the Collectivist (Vitalist) analytical perspective

To round off this description of the Collectivist mode of I-it identity, a summary follows of the important core valued concepts. Please remember that defined here in this section is the equivalent of one of an artist's three prime colours. This Collective perspective on human life has great value for developing humanity. However, without being balanced by the other two views it leaves people vulnerable to manipulation by others of ill intent. At its solitary extreme, this is a yes-and-no-no view of life. It can seem completely internally consistent, but only by excluding certain self-evident facts in the other two views, such as the endurance of matter as an equally logical assertion to the endurance of souls, and the presence of self-interest and *choice to relate* that informationally varies people's power over others. However, the Collectivist view, with its Participator mode of personal identity, is important to somehow integrate into a mysterious unity-with-diversity that includes *a sense of belonging in something bigger than oneself.* Without that integration, human life reduces to mere dust and stress.

The values or beliefs described here are for the Big Picture context. Their relevance in more local settings can be intuited and discussed.

- **Reality** is *behind* matter as a relational life force or esoteric energy that shapes the *illusion of matter* into forms that make visible the qualities and problems of the person's enduring soul.

- **The dualistic essence of life** is an *impersonal life force* (Chi, Hegelian spirit, Platonic Ideals, collective values) that shapes (or shape) the lives of all those collectively in a relevant space.
- **Space** is a context for eternal souls to interact, and may have several 'levels', or 'orders of existence', or 'heavens', depending on the qualities of relatedness that the souls at that level are able to sustain.
- **Time** is the eternal now emerging from eternity into personal awareness.
- **The dualistic personal identity** is a *Participator* in life.
- **Soul** is the person enduring through eternity at its various 'levels'.
- **Spirit** is relational connection between people.
- **Body** is a temporary form that makes visible the relational qualities of the soul.

Making yes-and-no-no judgements

How is it possible to integrate these three primary analytical perspectives on life with their vastly different and incompatible deductions? One way is storytelling or descriptive writing about the choices people make in their inner hearts as they balance up these possibilities and their priorities. The different analytical perspectives may be introduced to the listeners, or by their feedback comments and story extensions. By bearing in mind the summaries of each perspective at the end of the section above, we can develop a sensitivity to the unique features of each primary analytical perspective, and begin to enquire about the mixture and balance that may influence loss emotions and decisions made along the journey. The story has to be one that touches on values and chosen actions or rest, not only on feelings and behaviour that are individualised. Different people's perspectives on events can be added to the developing story picture, reframed to reflectively identify significant features from these three primary ones.

It is important to restate that describing the diversity of analyses here is not judgmental about their relative merits. Each has its benefits in different settings, and each its problems if it becomes a habitual way of thinking. When habitual, unbalanced values become self-fulfilling attractor states that limit adaptability and responsiveness. The aim is to increase sensitivity to hear these perspectives in the stories people tell of their life experiences, so that wiser choices can be made to sustain a life-enhancing balance of actions.

Telling brief 'stories' of life does not mean making up events, or telling lies to hide behind. It is a conversational way of living in which speakers and listeners identify *themselves* into the story and its values. I recall a car journey after a conference in which four people had more than two hours to fill on their way to an airport. A colleague suggested we each tell a two-minute story of some slightly challenging situation we had faced anytime in our lives where we had to make a choice. He started. The conversations flowed easily, with questions and comments and comparisons and curiosity growing by the minute until we lost track of time. The two hours passed as if in a moment. We had to exercise discipline and regulation to allow time for the fourth person to have her opportunity. An inner transformation, as well as a bonding into group identity, can emerge quickly if sensitivity and ubuntu allow an open relationality to bathe the material and individual facts of participants' lives.

In that particular experience people were not explicitly or intentionally reflecting on the presence or absence of the three primary analytical perspectives. It would be only a small step, however, to do so as part of a workshop or online event for example, where there is some prior teaching about these perspectives that could enable people to explore together their heart-level inclinations or bias for the choices they make when facing life's challenges.

Repatterning analytically framed beliefs into specifically named values is not so difficult once the possibility is entertained, and the informational purpose of loss emotions is understood to guide the naming process. A sense of awe and curiosity adds safety into hearing respectfully the unveiling of mysteries below the orientation horizons of our shared *life*. As people start to consider the implications of the new quantum gravity view, for example that life is *materially* shaped from within for us all, and via its social ecology is affected by our choices scaling into the outwardly shared world, then an important practical consequence will follow. To share our stories will become the most natural way to respond to changing environments.

The importance of timing when making choices

Three different types of time can be identified in these analytical perspectives on conversational life—linear time, timeframes, and the eternal now. None is more 'correct' than the others.

Triquetral principles of organisation describe movement as *the continuously changing relatedness of forms*. Movement cannot be broken down into parts. In a triquetral view of the cosmos, movement emerges from within, as life, or the potential for life's renewal.

When a personal choice is made, the direction of movement, of life, alters simultaneously in all three ecologies—inner, social, and physical. A personal choice will inevitably have progressive impact on life for all who share in those ecologies.

Given these considerations, the timing of a personal choice to alter active and stative states of the shared ecologies is important.

There *is* a relevant timeframe in which the decision needs to be made.

Within that timeframe there *is* an ongoing linear sequence of deterministic causal events that needs considering, which physically and materially impacts the progression of movement in the ecology from which all life-forms are continuously emergent, and in which the decision is being made.

Along that time sequence, there *is* a moment when an optimum impact of change would be set in motion that would contribute deterministically to moving the state of the whole system into a more life-enhancing mode. The measurable state of that shift occurs at the end of any relevant timeframe, so living with uncertainty during the timeframe is part of life-enhancement, although it may shift into a death-enhancing state instead if that is a consequence of people's choices. The moments of decision along the way within that timeframe are all personal critical points. Each one is a critical change process, best made within a potential gap in time that is chosen as a strategy set within the wider context of the person's whole lifetime and its eternal consequence.

Our choices, and the timing of our decisions for action or rest, initiate an ongoing turn in the story. We are all co-authors adding our signatures. Just as the Saan people, mentioned in the Introduction, have had to adjust in the last 200 years as the surrounding culture changed under the cumulative influence of Enlightenment Science and materialist attitudes, so now it is our turn.

Another brief practical exercise on how yes-and-no-no analysis works would be, for a short gap in time, to mentally step back to analyse the choices each of us personally could take sometime within the next 30 years. The purpose would be to identify life-enhancing choices that might influence the economy, and our three ecologies, as population shifts around the world and social disruption are associated with increasing extremes of climate. The exercise is in how to keep those decisions compassionate—so that life's triquetral movement can continue to enhance relatedness and synergise connection during times of change. Timing *is* important if people are to come through stronger and more able to state their values openly in the social ecology.

Chapter 9

Other Variants of Triquetral Diversity in Unity

The core diagram for the triune model of orientation in conversations, compassionate or otherwise, is the double triquetra, shown in Figure 5 and repeated here.

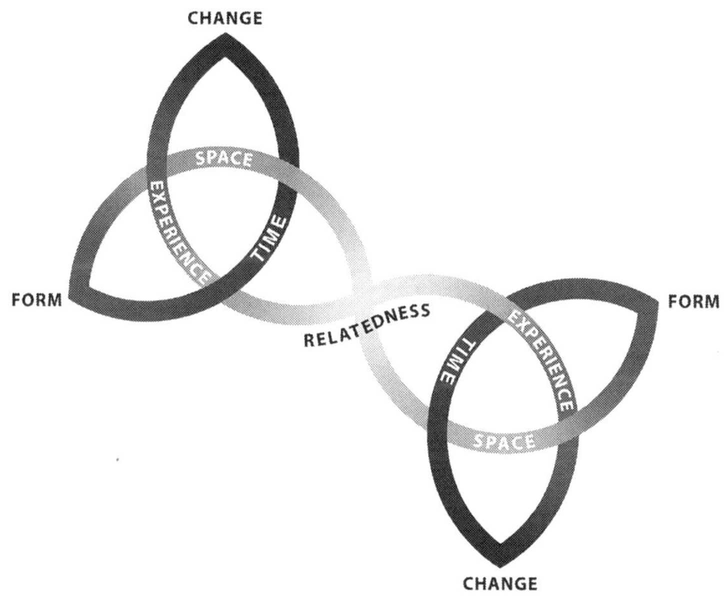

Figure 5: Conversational orientation in life

The three dualisms previously described all derive cerebrally from heart-level conversational synergy within wider ecologies. While analysis temporarily breaks conversational orientation, it can easily be re-established unless the analytical attitude has become a mental habit, and the values associated with it have become an attractor state. In this chapter, other variants and combinations of primary worldviews are described, all *non-judgmentally*. The aim of setting these descriptions in a triquetral framework is to increase people's sensitivity to the unity of conversational humanity that dynamically underlies diversification and subsequent coordination. Becoming curious with mutual respect opens the way to potentially growing in synergy while exploring how fruitful diverse beliefs may be, which is life-enhancing.

The previous teaching about recognising patterns of grieving is given to prevent the breakdown into divisions and exclusiveness of a healthy diversity. Grief emotions, when understood as parts of an integrated healthy adjustment process, are not barriers to relatedness even though when intense they may move people to withdraw for reflection for a time. They are the potential energy to become curious about, and to name, challenged values, then to explore how to reconnect with others by having conversations about them.

Shifting conversations from talk about feelings, behaviour, and possessions, to values and shared action plans is the practical outcome of triquetral synergy. Physical emotions are the corona of evidence pointing to the reality of personal values. The awareness of our and other people's values lights up the heart and mind with all the spectrum of shades of honour. Personal relationships are a physiology of light that shapes the molecular physiology of social biochemistry. Emotions connect. The thoughts of your heart can be renewed when choosing relatedness in your heart rather than choosing separation on a self-only path.

Monism

The Celtic triquetra displays iconically the unending dynamic of life behind all that seems constant. The double triquetra is the core icon of synergic relatedness, and of existence. It is noteworthy that the same three core features of movement that make neurobiological sense of how the triquetra constructs an orientation horizon—form, change, and relatedness sensory inputs—are all seen also in the Taoist taijitu icon (also known more commonly as the tai chi icon, or the yin-yang symbol) shown in Figure 21.

Figure 21: The tai chi icon—taijitu

In the West, many people are not aware that this icon is constantly on the move. If, for example, the dark yin 'half' of the circle expands, then simultaneously the light yang 'half' diminishes; except that within the enlarging dark yin is also being birthed its opposite in the enlarging white spot. This sustains equality without dominance or oppression, by mutual responsiveness and adaptability. As the light yang principle grows *within* the yin, it eventually overtakes its opposite in that location and becomes the more prominent feature of the circle, even while its opposite is birthed within, and so on endlessly. These core-level dynamics are seen widely in the psychology of relationships. In various timeframes and settings, men are commonly trying to connect with their inner female, for example, and women likewise commonly try to connect with their inner male, all while people are exploring how to remain true to their whole self.

The taijitu is an icon of monism and wholeness with a life flow within called chi. Diverse *qualities of movement* are shown in a dynamic of integration. The clue to this being a monistic picture of life at one extreme end of the triunity spectrum, rather than a dualistic one as previously described, is the fact that all three triune principles, including relatedness, are visible *inside* the uniting circle, as the mutual movement of the transforming yin and yang. Dualistic ideas iconically split the triquetra. The taijitu therefore reveals the triune principles of organisation within, in that the yin and yang features of chi's flow *change* in *form* as they *relationally move* and balance in every aspect of life. Many people mistakenly project their Vitalist dualism ideas onto chi, but this is not a necessary (essential) mental step. Yin and yang are counterpart features of the uniting movement called chi, not dualistically separated essences that optionally or mysteriously interact.[1] The counterparts of yin and yang in the triquetral model derive from, but are not one and the same as, the left and right spins of the quantum change.[2]

The taijitu displays the importance of unity of flow for practical living, by bringing the insight of relational balance and relaxation into the way people may exercise their power of choice during times of change. The double triquetra adds to this an iconic and memorable picture of how diversity can be welcomed in the source of synergy from within for ever-renewed life. Diversity in unity is essentially necessary for adaptable living during times of change, as we shall see more of in Part IV.

A difference between the monism of the taijitu, and the triunity of living triquetrally, is their capacity to differentiate the three types of time. If thought is limited only to monism, then there is no explanation or mental drive to question the direction or purpose of diversifying movement, either in life or in matter. The monistic view may lead to passivity. It therefore leaves people in peril of being drawn into the submissive attitudes associated with Vitalist dualism, which are basically loveless even though they may teach the importance of compassion. Love implies equality and mutuality amid emerging diversity. Love requires an intra-active choice, or even an effort of willingness, to sustain synergic resonance. Whether active and focused on a value, or stative and restfully at peace in relatedness, love is the core harmonic movement of life made manifest in compassion. The choice to be active around a named value, or at rest, must remain with the individual person if the dynamic of love is to be true relatedness and not a mere submissiveness. There is a condition of mutual responsiveness that may look like obedience or submission, but is in fact an active state of choice to sustain or recover the true balance and peace from which renewal of life and even creativity are born.

The triune dynamic transformations of relatedness shown in both the triquetra and the taijitu are pre-cultural and pre-religious. However, the taijitu was drawn in a cultural history where the cosmos was thought to have no timeframe, no beginning or end. The triquetra, on the other hand, introduces the notion of linearity and of timeframes into nature's and human movements, even possibly scaled up to cosmological levels. The taijitu shows a closed system; the triquetra shows an open system of life, both displaying identical principles of dynamic organisation within them. New Science is currently adapting the thinking of scientists to the evidence that the cosmos *is* an open system. Human society around the globe is just beginning to wake up to the implications of that. Understanding how the double triquetra is an icon of creative, synergic love may help when that love turns into its grief mode, so that the grief emotions do not drag people down, but can be harnessed rapidly into an exploration of how to sustain life-enhancing values that restore love's relatedness to its joy mode.

People in a culture shaped by Collectivist values that have become associated with a dualist Vitalism may feel the need to turn to divinatory methods to interpret the signs for discerning the complex movement of yin and yang in life's physical events. By seeing patterns within physical, material events in small scale, many over millennia have believed that they can predict life's future potentials, as if the ever-present now is a holographic representation of the whole. The *I Ching* is a divinatory book—the Book of Changes—developed by Chinese sages over the last ten centuries that guides people to interpret patterns by casting sticks, stones, or coins. This need for divination into the fractal scaling down and up of changes within a monistic essence is the clue to the optional essence-mind dualism that may remain hidden for some within monism.

Observers of life in a Materialist mindframe would probably consider these divinatory fall-out patterns to be randomly predetermined by conditions in the caster's hands. Creators of life in an Individualist mindframe would say that neurological patterning about hidden needs or wants would guide the interpretation, more than any possible fractal or holographic scaling of the physical ecology. Participators in a Vitalist view of life might see reflections of eternity in the visible patterns. Truth is personal, not found in an essence. This is why the double triquetra may show an underlying truth about synergy, and its further scaling, over which personal choice has a deterministic effect within a systemic 'transferability' logic. Seeking a balance, between respecting the whole and respecting the local physical constraint of matter, is a *conversational process*. That process has a final cause, which is to seek the critical point of balance in relatedness deeper than seemingly incompatible views and beliefs. Life-enhancing choices could follow in that process of seeking, if conversational responsiveness is held sacred at every ecological level of human being. If, on the other hand, that focus of attention shifts from synergic balance to only inner self-purification, or to oppressing others into conformity, then the potential for the renewal of life remains at a distance, always in waiting, rather than finding its fulfilment in the eternal now.

Tripartite thinking

Taking a timeline in the Middle East from its ancient Chaldean past five millennia ago to the present day, the family of Abram, later renamed Abraham, has diversified into many nations, and into three monotheistic religions—Judaism, Christianity, and Islam. This migrant family has developed its own continuously diversifying and adapting closed groupings, each setting different relational conditions for individuals and peoples to become open systems approaching the same perceived One creative source of life. Each holds to a notion of unity among its community of people, this unity of social ecology being a central focus of their spiritual life. However, each also sets boundaries with exclusiveness, which has generated millennia of violent conflict within the family. Each *umma,* communion, or nation admits a range of views about how much diversity is welcomed, and the grounds on which others should be excluded.

The source of life for this family and its adopted members is an Almighty Creator, common to all three religions. It is not widely known that this is the Chaldean god called El, who Abram believed called him to *move* with his family and flocks. This Chaldean root is found, for example, in the Arabic name for the Almighty, Allah, which is a poetic contraction of *al-El-ah,* meaning '[the]-[Almighty One]-[exclamation of awe]'. The point being made here is that the cultural history of this extended family reintroduces a relational quality *among people* into its concept of wholeness and perfection, in contrast to the impersonal concept of a life force coursing through space and a collective of individuals. I say reintroduces, because this relational quality has been lost, as described in the mythical story of Adam's fall—*Adamah* meaning both earth and humankind including male and female equally. The story of Abram's journey with family and flocks to restore that relational quality, which continues to this day as Abraham's family journey, connects people and the land in a shared ecology (over which there is much conflict). The land nurtures humanity into growth, and humanity likewise nurtures the land to be fruitful. This is an ecological spirituality of synergy. Physical, social, and inner ecologies all meet in the ongoing story of hope for life to be restored.

But that quality of relatedness *has been lost*. It is still lost, and the religious teaching in all three monotheistic religions is that it needs recovering. The resulting separation and brokenness between people and peoples, and between humanity and the land, has fractally replicated down through millennia, both in global history and in the way many family members think. The notion of overcoming separateness as an explanation for suffering, and to mark out a path to reconciliation and prosperity, has become an attractor-state value focused on recognising separation.

This attractor state of hope for unity and its life-enhancing impact has had an unfortunate consequence. For 3000 years of recorded history, intelligent people have taken an overly analytical approach to understanding the human condition and the nature of life, resulting in a tripartite notion of the human being in philosophical and theological thought, as contrasted with triune.

Figure 22 shows the resulting idea of a 'holistic person', complete in all its parts, which are separable body, soul, and spirit. To say that the whole is greater than the sum of its parts gives no explanation for how that can be, and therefore gives no guidance on the way to enhance life; or at best it produces conflicting guidance. In an Enlightenment mindset this notion is more commonly described as *body, mind, and spirit,* but the value of 'separateness overcome in some way' has pre-filtered sensory feedback from the lived environments in this tripartite way of thinking. This pre-filtering limits the capacity to break out of this closed-system way of parts analysis. Prior separateness is assumed. The value system that sets boundaries around parts has even been applied in some Neo-Freudian psychology to id, ego, and superego being considered parts of the psyche. A healthier and clearer way is to see them as perspectives on one integrated movement of systemic life in a process of renewal from moment to moment.

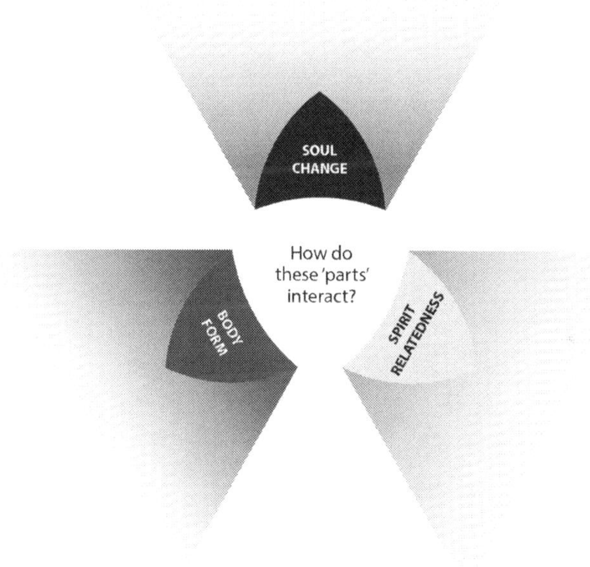

Figure 22: The tripartite view of a holistic person as separable

Tripartite thinking about an individual's inner nature is in contrast to the social double-triquetra *triune view* of a whole person in conversationally shared life (the ubuntu of belonging in wholeness). The contrast becomes significant for understanding how renewal of life can emerge *from within* the whole, a systemic whole, rather than by imposing an externally applied law or an action on individuals who are perceived to be separated from the active agent. The law of love is at work below the orientation horizon, however, even with those who are mentally separated from each other, while above that horizon the law of behavioural control imposes its own type of 'conforming order'.

A further cognitive error that follows from pre-filtered separateness is to extend the tripartite image of Figure 22 into the Venn diagram shown in Figure 23. Each 'part' is represented as a self-existent concept, boundaried as a circle. When these overlap, they generate a triquetral shape in the centre, which is used to recognise a range of overlap states of interaction between these separated elements.

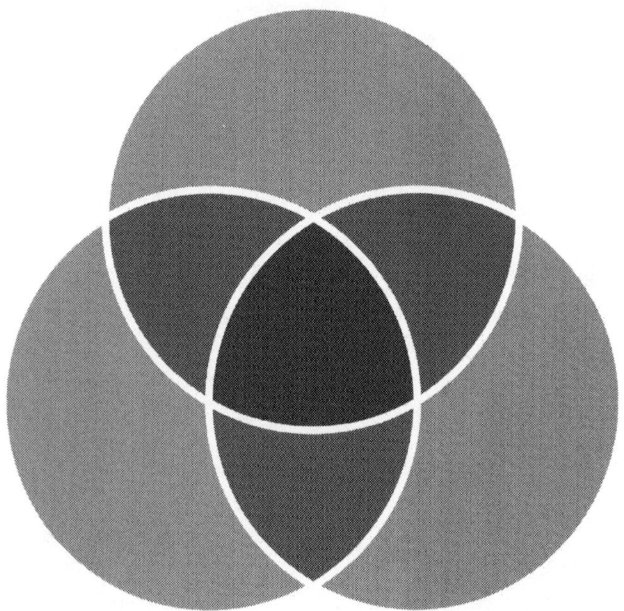

Figure 23: The Venn diagram error of overlapping separateness

This is diametrically opposite to the triune notion of a primary integrated *movement of life*. The Venn diagram would suggest that *life* is an optional product of interactions between three uniquely different and self-existent essences or entities. Triune thinking contrasts with this. Triunity, as shown in the Celtic triquetra, asserts that eternal movement, as the informenergy we know as life, precedes all that we cognitively construct in our pre-filtered minds. In triune thinking, the Venn diagram circles are only figments of our imaginations. They reveal the brokenness in our hearts. The truth is personal, as a chosen responsiveness in our adaptive relatedness with each other, extending also possibly to a uniting source of life whose informenergy extends way beyond ourselves. This truth invites us into a renewing relationship of love, shown in the double triquetra.[3]

Each of the three dualistic reductions from a triune view is a yes-and-no-no picture of reality. Trying philosophically to shake these three together into a spurious unity of parts moves life in the wrong psychological direction. It would remain cerebral and broken, a juggling act in the higher brain's feedback loops that leaves the inner heart without words to express its intuition that something greater is missing from this way of seeing life. Tripartite analysis cannot reach deep enough to connect a person's inner heart dynamically into living networks of relatedness with conscious awareness partnering reason into the power of choice. There has to be a no-no-no response to tripartite holism.

Creating no-and-yes-yes political comfort zones

Having described yes-and-no-no dualist views of life, and a no-no-no tripartite view, a brief look at how people create political comfort zones in which they can ignore all this time-wasting philosophy will be time well spent.

Unfortunately, most people make comfort zones by excluding something they dislike. The same is true with excluding unpredictability, which can be done by simplifying fully triquetral living. When one of the three orientating foundational principles is not very appealing, omitting it from a view of the world and oneself increases the predictability arising from anticipation patterns, and increases pleasure. Avoidance thus adds a sense of security and the empowerment by control of a dangerous world. People then can imaginatively create *no-and-yes-yes* views of the ideal life for 'human beings like me'. By sharing this set of beliefs with others, people can increase their security further, until eventually they may try imposing that set of beliefs about a reduced view of life on others. This is called political vision, and it creates political realities instead of religious, or philosophical, or plain human or family ones.

To clear a way forward from restrictive religious, political, and philosophical realities or visions, the life process of inner renewal could be understood in triquetral terms of *security* at every level of ecology—physical, social, and inner—to transfigure life from within. This involves teaching how unpleasant loss emotions are the active grief mode of enduring love, generating heart-level movement to restore connection and adaptability. They are not signs of weakness or dishonour, as is misunderstood in many overly analytical and overly regulated cultures. Violent conflict arising from loss emotions does not reduce grief. That choice fractally scales grief into a fission-type emission of informenergy into chaos, leaving many nasty products of degradation to pollute future life. By spreading triquetral understanding, with the double triquetra as the core icon of synergic relatedness, the opposite process of fusion may release informenergy into the potential for renewal of life instead. Without degradation, fusion encourages social transformation. The realistic hope is to become secure enough at heart to allow diversity in our space and timeframed experiential presence of life with each other. Small groups and large can exercise creating gaps in time together, in which to integrate that which has been separated.

No-and-yes-yes—don't like relatedness

When people are not convinced that relatedness or mutual responsiveness is foundational to life, perhaps if they have had unpleasant experiences of relationships earlier on in life, they may seek to create a world in which *form* and *change* are their reliable elements. We need to look again at Figure 20, where relatedness is projected out from the person, but now in a very different light.

The reality of, and authority over, a distasteful external life is now internal, in the individual's identity. It is as if the scales have shifted, from belief in the projected reality of an external higher power or a community that has power over one's self, to self-belief in the power of the inner life to erect real barriers that keep relatedness out. It is Individualism turbocharged with Materialism.

Predictable time in the external measurable world may help feelings of control over it, and build safety within the self. Therefore, objective time, where time sequence and timeframes fuse, may be an attractor-state value that pre-selects sensory inputs for analysis. The ego circle is then strengthened in this comfort zone as a *personal reality* that brings together the powers to regulate form and change in one's own timeframes. Now Materialist-Individualist individuals can agree that there is a real and measurable physical world in which *subjective time* is also a reality, and that subjectivity is more important than the broken physical world out there.

The mystery of how objective and subjective time connect or disconnect is less interesting than enjoying the subjectivity and comfort of this combined worldview, where external criticism can be shunned as an uninformed intrusion into personal space. In this combined worldview, spiritual beliefs are considered subjective and internal, so a wide range of ideas about eternity, life, and death can be entertained without disrupting the security of a predictable world framed in Enlightenment Science. However, the New Science does challenge the notion of objective and subjective divides.

To take a risk to think outside this mindframe, consider that neuroscience demonstrates the power of choice resides in the limbic brain, at an emotional level neurologically balancing cerebral anticipations. There, in the inner heart of one's personal values, the self and the world truly meet preverbally in mutual feedback responsiveness. Naming one's personal values firmly at heart is a more life-enhancing way to resist external power than erecting and then shoring up barricades.

Saying 'no' to relatedness, and 'yes-yes' to form and change, primes an individual to resist an imposed authority. Individual assent for external change must come from within. This creates the psychological foundation for a political system of democracy with individual rights, but sadly without a balancing belief in matching duties to contribute.

No-and-yes-yes—don't like change

When people are not convinced that change adds any value to life, for example if they have had unpleasant experiences of living in chaos, they may seek a world in which *form* and *relatedness* are their reliable elements. Figure 19 now reappears, but in a very different light by which the scales that determine reality have tipped, away from the stressful external world where the individual is under the authority of externally projected *time*, towards the inner self as an island of peace, stability, and good order.

As a result, the idea that the Materialist world is predictable in the relatedness of its forms may be comforting. This structural framework can comfortably partner with a Collectivist sense of relatedness, for example belonging as a member of a tribe or community or family that has a recognisable geographical origin, with recognisable physical rituals and traditions that survive even in a diaspora. When Materialist and Collectivist perspectives seem to fit well together, Individualist ideas about innovative change and diversifying creativity may seem out of place.

The belief shared about the external world by Materialist-Collectivist thinkers is that it is spatial. There is a real space or territory in which I belong with others like me, which we together can shape the way we like it to preserve our comfort and secure our predictability. To have a garden gate that can be closed on a family home, or protected borders to our nation, or established behavioural traditions to our religion and place of worship, helps to define these physical territorial boundaries. Individuals introducing change who cross these borders may shake that sense of collective security.

The collective authority within this territory will therefore resist change, opening the way for a small ruling elite to declare and impose standards that a population must accept for their security. This can lead to highly structured traditional societies such as in Korea and China, but also to a justification to use the increasingly structured knowledge of material science for the surveillance of individual lives to bring them into that social order.

As seen throughout the world's history, people who retain individualistic or innovative ideas might react to any corruption of that central power with social unrest, which would lead to a crackdown by those who prefer to remain in this limited perspective on human potential. This leads to a majority in such a society preferring the political system of dictatorship.

The counterbalance to excessive oppression that stifles innovation is the New Science notion of adaptability. Emergent order can be re-established in changing environmental conditions by responsive feedback learning from those who differ in their approach to life. The synergy of informenergy in mutually informed responsiveness offers another way to disperse the fear that change will remove security. The New Science affirms that true security for life emerges from within the mutuality of transformations.

No-and-yes-yes—don't like form

When people are not convinced that form or rigid structure is foundational to life, for example if they have had unpleasant experiences of entrapment in a restrictive system, they may seek a world in which *change* and *relatedness* are the reliable elements. Figure 18 becomes relevant again, but in a very different light. The scales that determine reality's values have tipped, away from objectivity, towards a firm belief that to live in one's own subjective states of being is the whole point of living, making whatever they want of life as an individual, and no-one has the right to criticise them. This intense focus on the Individualist as reality can curiously become comfortably mixed with wanting to flock together with like-minded people, to fly in freedom as they will. Thus, Individualist *and* Collectivist perspectives can surprisingly fulfil a person's heartfelt desires, and Materialist ideas of ordered structure and determinism then seem out of place or irrelevant to life.

The belief shared about the external world by jointly Individualist-Collectivist thinkers is that it contains the presence of 'me' as an objective something of significance, a soul perhaps, albeit embedded in some bigger context that feeds me from within. Here we see the reasoning behind the ways that Eastern Collectivist philosophies have entered Western Individualist societies to create New Age thinking. There is also the reverse social process, the breakdown of Eastern traditional society when individuals travel to the West to advance their educational achievements and discover Individualism.

This doubt about the value of traditional structures to society or life leads individuals to resist the imposition of external pressure or obligation to conform. The preferred political system is anarchy. This can create mayhem in educational systems, for example, where a culture of challenging the pass criteria imposed by exam boards reduces the capacity for formative comparisons of skills and knowledge. Minority interest groups that apply pressure within any political and social structure are another example of the wilful disruption of form by gatherings of individuals.

Infinitely diverse variants of triquetral unity

Just as an artist can mix three primary colours to make an infinitely nuanced picture of life, so too every human being has that capacity for uniqueness. The three primary analytical reductions of conversational orientation into I-it modes are summarised here. The purpose is to ease comparisons in a way that may stimulate conversations with those who see life differently, thereby seeking ways to integrate them around named personal values that enhance life.

- **Reality**
  —*Structuralist (Materialist)*: Reality is matter
  —*Individualist (Informationist)*: Reality is process
  —*Collectivist (Vitalist)*: Reality is connection
- **The dualistic essence of life**
  —*Structuralist (Materialist)*: Energy-mass understood as wave-particles
  —*Individualist (Informationist)*: Information packaged to repattern life
  —*Collectivist (Vitalist)*: An impersonal life-force
- **Space**
  —*Structuralist (Materialist)*: A container in which matter has its own momentum

—*Individualist (Informationist)*: My personal mental bubble interacting with other people's personal bubbles
—*Collectivist (Vitalist)*: A context for eternal souls to interact, which may have several 'levels' or 'orders of existence'

- **Time**
  —*Structuralist (Materialist)*: A linear process (the arrow of time) as a causal sequence from past to future only
  —*Individualist (Informationist)*: A stressing timeframe within which changes must be made to create or fill life
  —*Collectivist (Vitalist)*: The eternal now emerging from eternity into personal awareness
- **The dualistic personal identity**
  —*Structuralist (Materialist)*: An Observer of life
  —*Individualist (Informationist)*: A Creator of life
  —*Collectivist (Vitalist)*: A Participator in life
- **Soul**
  —*Structuralist (Materialist)*: An incomprehensible term, for which 'mind' seems a fair substitute
  —*Individualist (Informationist)*: The activities within a personal bubble that generate information
  —*Collectivist (Vitalist)*: The person enduring through eternity at its various 'levels'
- **Spirit**
  —*Structuralist (Materialist)*: A mysterious and unmeasurable concept, best considered an archaic mode of primitive thinking that has been superseded by our deep knowledge of the physical structure of matter and energy
  —*Individualist (Informationist)*: Informational transmission from the personal bubble
  —*Collectivist (Vitalist)*: Relational connection between people or souls

- **Body**
  —*Structuralist (Materialist)*: Self-organising matter under the influence of DNA and RNA
  —*Individualist (Informationist)*: The shaping into a form or pattern of any substrate that carries information
  —*Collectivist (Vitalist)*: A temporary form making visible the relational qualities of the soul

Chapter 10

Speaking Life

Triquetral yes-yes-yes balance and curiosity

Finding a yes-yes-yes triune balance in one's approach to life is not as difficult as the foregoing discussion may suggest. To be actively on the search is sufficient prerequisite. Triune curiosity does not negate the value of diverse philosophical, religious, and political relational systems—quite the opposite is true. Triune curiosity reduces the need for conflict between them. It shifts the conversational focus from beliefs to naming the underlying personal values that connect with others or diversify from others. It shifts the emotional preparations to respond, from reactive and defensive behaviours to planned action that preserves named values agreed with people who see other features of life differently from you. This focus of attention on personal values and agreed actions to preserve them may enable reassessment of the significance of past events, as well as preparing for a shared future. Forgiveness or acceptance and moving on become realistic possibilities when speaking life triquetrally to each other.

Triune balance does not mean that people should avoid analytical thinking, or that analysis is somehow morally wrong or inferior to 'being in the conversation'. Times of analysis in all three foundational perspectives are vital to making *informed choices* about how to respond to environmental challenges. Understanding the triquetral balance simply empowers people to know more insightfully that they are being analytical. With that comes the need to decide when to restore a conversational mode with others who see life differently. The conversation moves to compare, contrast, and potentially integrate mindfully the 'outlandish ideas' that will have been generated. For example, a person may intelligently adopt an Individualist attitude while at work to get through the stressful task list, then swap into a Materialist mindframe when the car will not start to get home, and resign into a Collectivist mode when calling for help from the recovery services, before finally getting home and relaxing into a conversational balance as they tell their story of the day to friends or family with as much good humour as they can muster organisationally in their informenergy while they raise the synergy of fusion by sharing the drama.

The analytical exploration in each of these phases of the story will have been stretching one of the three corners of the triquetra outwards to focus attention on a new perspective, before relaxing it back with an emotive adjustment process as each perspective's angle on life is let go while we explore restoring shared conversational humanity to tell the story. The Emotional Logic way to understand emotional processing enables any loss emotions thus generated to empower letting go, followed by choosing to explore something new in this mutual adjustment process based on the named values challenged in the situation. This release of emotional informenergy adds a profoundly heart-level strengthening to the ongoing fusion of synergy as people hear and respond to the story. With heart and mind in partnership, the light of consciousness shines not on the self during storytelling as much as on the wider ecologies in which the self belongs and moves with others. Generating synergy in these feedback loops extending out from the body is *speaking life*.

Information (verb) creation in the human brain

Part III of this book unveils the mental tricks hiding a shared informenergy reality that constantly moves synergically. In Part III we have cleared a way through the inner ecology, so that triquetral movement may be seen informationally everywhere and for all time. In Part IV we shall go on to clear that way further out, through the traces left in the physical and social ecologies by the impact of past analytical thinking and its associated anticipations. An ethical balance emerges that can guide us in how to speak life, choosing not to speak deathliness any longer in the stories we tell. Speaking life is not so difficult as to be beyond anyone. To round off Part III now, we return to the inner ecology. We shall look at some practical ways to sustain triquetral understanding in a world that is now substantially misshapen by the over-analysis and self-protective fear that has recurred cyclically over millennia.

In the human brain, and in other species, feedback loops are synergic. Cyclical informational processes (informenergy) generate resonance between remote localities. This informenergy movement simultaneously generates substantial patterns and mindful understanding. However, these two can dissociate into separated ecologies, as we have seen. The synergic processes in the brain's neural networks may drift off into their own internal neurophysiology, divorced from the rest of life. As we shall go on to explore, crossing the fuzzy boundary between inner and outer ecologies is the molecular-scale physiology that has been very well mapped and named by Enlightenment Science methods. Now, using New Science methods, there is also an *informenergy physiology*. Emotional Logic reflective tools can help to map these dynamic patterns of ecologies wider than biochemistry, showing how they influence its probabilities for change. Molecular physiology is embedded in informenergy physiology. Molecular physiology is not separate to the informenergy physiology, or parallel to it, but emergent from it *and feeding back into it*. One is not more 'real' than the other. They are different levels of order-shaping according to identical principles of organisation on vastly different scales. Each will have its own fractal range of activity-patterning, which may appear to process uninfluenced by the other level. However, they will mutually exert probabilistic changes across the scales, as when the earth's rotation affects the way gravity pulls water down through a plughole.

Therefore, neuroanatomical feedback loops in the human brain not only shape the well-known molecular physiology, but also they create resonant mutuality in the quantum informenergy physiology, as well as scaling up into the social physiology. Double triquetrae of feedback could be imagined extending within the molecular physiology of the brain and body *and beyond it* to picture how principles of organisation create human mindful intra-active movement. This is the informenergy environment for speaking life.

Measured comparable proof of this informenergy extension beyond molecular physiology is seen in biophysics. Electromagnetic fields around and within macromolecules such as DNA, and across cell membranes, can be measured. These radiate their influence around the 'object' to shape probability fields for measurement events. (Particles when measured are mutual change events. They are *not* static lumps that randomly bounce around.) Some of the attempts to describe neuro-informenergy physiology have appeared also in storytelling ways, discounted by Enlightenment Science because they are not framed in a measurable Materialist perspective. The notion of 'humours' moving around the body along meridians is one such way of framing the empirical discoveries of intelligent physicians in ancient Chinese culture, for example. These 'humours' do not need to be mentally conceived as ontological life-forces, however. The taijitu is an icon of flow, as is the double triquetra, with no prescriptive requirement to think ontologically when people tell their empirical story of the synergic changes they have noticed when acupuncture needles are inserted. The Chinese culture of its day did not allow one person to touch another, considering such contact shameful unless married. Ladies in particular used model manikins to point to where their symptoms were. The only means available to physicians to intervene was to apply pressure with needles in their well-intentioned attempts to alleviate distress. It was in a physician's interest to do so, because the economic system of that culture was one in which people paid their physicians regularly while they remained well, but stopped the payments when they were ill until the physician had restored their wellbeing. Doctors were bound to explore how to restore their income intra-actively.

Biophysics is more 'dispersed' now that New Science has made *systemic* sense of emergent order from chaos and complexity. Wave-particle non-locality of quantum processing *is* determined in its outcomes by environments that include the changes induced by an observer. The iconic image of iron filings aligning on a sheet of paper as a magnet is moved underneath is now overtaken in the mature mind by MRI scans of the brain seen commonly in social media—*magnetic resonance imaging*. The technology behind this is awesome, and quantum. A powerful magnetic field around the head or body induces every proton in all the atomic nuclei to align their spinning, held in the magnetic field by synergistically responding to its informenergy input. When the field is modulated by a pulsing radiofrequency wave from different angles, which may happen several times a second, every molecule additionally tenses its spin and then reverts to its former alignment spin, and in so doing emits the excess field informenergy it had received. The amount of tensioned emission from each proton depends on the surrounding environment of those atoms and molecules (a vital point to take note of, to be recalled in Part IV), such that the water density in various body tissues can be mapped and turned into the scan images that are now so familiar. Thus, several scales of order are involved in MRI scanning, and yet the powerful magnetic field does not alter the molecular physiology of the person being scanned. Scales can function seemingly independently, but there are specific resonances that do cross them. For example, bone fractures heal more rapidly when a small electric current is set up across them. Cartilage in joints can regrow when a small piezoelectric mesh is inserted into an arthritic joint. Even more scales of ordering of informenergy are involved in the decisions subsequently made by physicians interpreting MRI scans, who now absorb the informenergy shaped into the scan images and respond synergistically by choosing to informationally activate selected items from a vast array of intervention possibilities to benefit the patient in their lived social ecology. Most of the interventions have been made available safely as benefits of Enlightenment Science.

The aim in speaking life is to be *as* curiously inclusive in preparing to speak as a triquetral balance will rationally and emotively allow, but also to be curious intuitively about the character and purpose of a storyteller.

All of this is process. None of this conventional science Materialistic storytelling approach to healthcare needs to contradict a parallel story being heard by people who are suffering; such individuals may choose to hear from others who, for example, have benefited from Chinese acupuncture in a Vitalist storytelling frame of mind, or from those who have adopted strict attitudinal self-regulation and diets in an Individualist storytelling frame of mind to boost their immune response to disease and its associated experiential illness. Here we can carefully apply a yes-and-no-no triquetral analysis of balance. Each of these perspectives on healthcare is framed analytically in ways that will be deductively incompatible if pushed to an ontological extreme in habits of thinking. However, they all find their unity and potential synergy for social health and wellbeing in an underlying triquetral process of emergence and dispersal. Each perspective informationally gathers a feature of emergent order. Then, when also letting go of it into a triquetral conversational mystery, any grief emotion that holds the tensioning as extra informenergy (like a proton in a magnetic field) can empower moving life on to renewal with new insights gained.

To speak life into health promotion and wellbeing by applying a yes-and-no-no analysis, and thus avoiding speaking deathliness, we shall need to look again at the earliest origins of Enlightenment Science. In Chapter 6 I described how King Charles II issued a Royal Charter to found the Royal Society of London, "whose studies are to be applied to further promoting *by the authority of experiments* the sciences of natural things and of useful arts". Why did he sign this Charter on 28 November 1660, in the first year of his reign (1660–85)? Historians obviously debate this, but one answer may be his desire to prevent charlatans from manipulating people by claiming supernatural powers to heal or influence their lives or relationships. The reason could not be the twentieth-century modernist crusade to eradicate superstition from society, because Charles II himself believed in the royal prerogative of divine healing by the 'Royal Touch'. For centuries, British and French monarchs had this divine right attributed to their status. Charles II exercised this healing therapeutic touch more than any other British monarch, especially for a skin condition known at that time as scrofula. However, he was also very much a realist about deception, having survived the Cromwellian republican revolution and been recalled to restore balance to a society that had been torn apart by claims of divine right. He knew that simple people were easily deceived by those who made claims of a higher authority or knowledge. He needed to unite a divided kingdom, and the Charter would appeal to realists everywhere. This noble purpose continues to this day to motivate the search for scientific proof of claims for authority and knowledge. However, a strange set of circumstances has evolved over the following 350 years. Science bred the 'useful art' of technology. Technology bred sales, which bred an economic revolution resulting in alienation from the land and from human values. As trust in a divine and providential creator and healer dispersed, so Materialistic beliefs gained sharper focus and increasingly specialised knowledge. People became enslaved to the resulting hope for economic progress and luxury, while political ideologies based in the Materialistic perspective shaped living systems. Exclusive Materialism has become its own charlatan if its knowledge is used to deceive an unwary marketplace.

But the original purpose was, and remains, noble. All that has been lost is the rational capacity to balance the claims of Materialists with those of Individualists and Collectivists to secure future choices based on shared values for humanity in balance with its environment. If any primary perspective becomes weakened, then charlatans may try to profit from exaggerating the claims of one or two other perspectives. The ability to dispel dualistic assertions of absolute powers and control is important. Yes-and-no-no analysis introduces a triquetral authority of *conversational balance* as the truth criterion for confidence to act, or to rest as appropriate for situational changes, or to resist manipulative influence by stating human values. Seeking that way can alert attention to potential danger when someone claims authority in one perspective or another. Asserting the need to join in the diversity of informational flow builds a stable domain. Then responsive adaptability generates informenergy renewal in groups and systems, enabling them to survive and thrive in changing life-circumstances. Speaking life has a firm foundation in the truth criterion of transferable conversational balance.

Information (verb) loss in the human brain

The neuro-informenergy of feedback loops between the cerebral cortex and the limbic core brain (of which the thalamo-cortical sensory feedback loops have been most studied) may also recruit 'horizontally' to connect neighbouring or remote loops into coordinated harmonic or resonant bundles. Bundling produces the neuroplasticity involved in learning and attention focusing in social settings. Informenergy movement thus feeds across the whole inner ecology's self-organisation of biochemistry, biophysics, and informational processing. As attention changes its focus, so the entire pattern of brain functioning repatterns with it, and with that the body physiology and its multiple social messages.

Adaptability to changing external circumstances, however, requires also the dispersal of some former patterns of inner ecology as attention shifts. Information builds up in feedback loops, not as an accumulating noun, but as an iconic verb understood as a *double triquetra repatterning* of resonant informenergy at vastly varying scales depending upon the focus of attention. Simultaneously, informational shift of the entire person's disposition generates a new emergent pattern along with 'information loss' of the old. This is a process view of adjusting mental focus being manifest in adaptive social behaviour patterns. In this we may see the taijitu in movement. As one pattern of understanding grows, another disperses, unless it is held in memory. For example, focusing attention in a conversation on naming and agreeing personal values requires that 'surround inhibition' by inhibitory neural feedback loops enables previous thoughts and preparations for action to disperse to concentrate on the task. As previously described in Chapter 3, surround inhibition of neural activation sharpens the informational content of interesting sensory features. Simultaneously, it screens out distractions from a previously chosen purposeful action.

This screening-out process may be seen to normalise information loss in informenergy processes. The presumed problem in Enlightenment physics of the conservation of matter, energy, and information is myth, associated with a closed-system pattern of thinking that generates analytical dualism. However, care is needed here not to merely swing across to a different dualistic reduction of life as if it is a solution. The taijitu is iconic of triquetral movement, although potentially imagined to be a closed system where there is no energy transfer in or out. Reflecting on it may unfortunately push the unwary towards a Vitalist dualism to sustain openness in life. The double triquetra is iconically safer as a mental focus of attention to keep the principles of organisation at heart. In each triquetra, personal values move the inner understanding of life towards an open conversation, which then synergistically takes substantial order into the world, which may then convert it into progressive energy to change wider systems. The human being is not merely a passive recipient of life. We each are synergically intra-active recipients *and givers* or speakers of life.

The brain's highly complex neural networks are embedded in an open system that innately receives informenergy input through our embodiment. Simultaneously, we output movement informationally into wider ecologies with which we may discover resonances that enhance life, even beyond our local space-and-time frame of living. In this synergic physiology, neuroplasticity creates fuzzy boundaries of renewal and dispersal patterns of informenergy. The brain is not a closed system. Information (verb) is both created and lost in this relational process, a characteristic of open systems.

Sadly, surround inhibition and screening out of information (verb) that is not currently of interest to the individual happens even when the focus of attention is on death-enhancing values. These closed-system values tend to follow dissociation of conversational synergy into one dualistic pattern of thinking or another. All the dualisms introduce a potential for attitudes of manipulation and control to creep into anticipations. The more is invested emotionally in preserving any such ego state by screening out options, the more difficult it becomes to change direction in life. Loss seems increasingly inevitable if a choice is considered to change perspective on life. Therefore, learning to understand and befriend unpleasant loss emotions is important, enabling them to be understood as the emergence of human values-based informenergy. Exploring choice options with them connects the inner heart into stable domains of resonant double triquetrae beyond the individual. These dynamic connections sustain an individual's life during transition to a new personal state while letting go of the old. This paradigm shift into personal extension is needed to explore how love returns to its relational joy mode. Having a mental iconic picture of stable domains can make the choice more reasonable to transition from analysis into conversation. Part IV will contribute to this iconic picture.

The other workable solution to prevent a self-fulfilling crisis of deathliness from accumulating is to reduce *cerebral* activation by opening heart-level *awareness* to a renewing, life-enhancing informational input. Fortunately, the neurology of the whole brain has systems in place that enable this to happen. In the brainstem are the regulating centres for breathing and pulse rate. These can disconnect from each other when intensely emotional responses radiate as a corona from sensory inputs. Adjacent to these regulating centres is the *ascending reticular activating system*. These nerve fibres ascend without synapses to prime the surface of the cerebral cortex to receive new informational inputs that will be delayed through the synaptic filters. By choosing to focus, during a brief gap in time, on matching breathing to pulse rate, a side effect is that cerebral activation reduces. The analytical focus previously holding the brain in its attractor state can then more easily disperse. The core brain's filters are released to reset into an awareness, below the orientation horizon, of belonging in life-sustaining ecologies. Body posture will then relax, in the course of which embodied memories locked into posture or poise are also dispersed. The whole person, now aware of wider ecological connection and released from habitual analytical ways of reasoning, is liberated into a new potential for renewed life. The contemplative state must give way again, however, as in a taijitu movement to explore socialised activity in the wider ecologies. If the inner ecology does not scale fractally into socialised behaviour, inner dissociation will re-establish itself by the lack of social feedback awareness.

Therefore, choosing a gap in time to reflect can open a triquetrally balanced inner state. In this state, unexpected mismatches in the core brain's filters will have greater surprise value (informational value) than targeted anticipations for which there is a successful match. Acknowledging this feature of New Science shows how triquetral reflection in a gap in time differs from the type of meditational practice that imagines a state of inner purification in which all desires are eliminated in the flow of an esoteric vital energy. That type of meditational practice may not lead to a triquetral view of balance. To be present triquetrally in a loving collective is a synergic inner state. This stative state does not require self-emptying so much as self-balancing one's own qualities of relatedness with those whose embodiment differs. In a mindful search for triquetrally balanced relational life, a little bit of randomness can stimulate understanding simultaneously with substantial transformation of life's values. This could lead to physical repatterning of life and wise risk-taking to converse with others.[1] Loving the renewal of life that synergistically follows is a more life-enhancing path to follow than mentally dissociating and separating from the world to live only in an informenergy physiology. Conversational curiosity adds to the life-enhancing path that others can synergise along. Encouragement thus received by synergy alters the filters. Re-filtering lights up the mind with renewing potentials. New options for informational choices are thus created, which can steer life's action better in balance with others through times of unpredictability.

All of the inner ecology moves with informenergy within wider ecologies. In the eternity of the cosmos, our local solar system is a *physical* patterning of informenergy that innately has become life-enabling on the surface of our planet. Millennia of annual rotations of our planet around the sun have enabled a *social* ecology to emerge, of diversified self-organisation of vegetation and organisms. Over a human being's lifetime on this planet that same relational informenergy emerges in the *inner* ecology as mindful choices of behaviour, which empowers life-enhancement, or corrupts into death-enhancing dissociations.

Triquetral Systems In Parallel EXchange: the TSIPEX model of conscious mind

Comparative consciousness is generated in the inner ecology from responsive awareness as two feedback systems mix. Through embodiment, the social and physical ecologies bring sensory input to the core brain, where it impacts the inner ecology. Both are triquetral systems, creating a double triquetra iconically within the self. When these two parallel systems dissociate, they can reintegrate by choice. This is an informational model of *movement* that unites matter and mind. We become embedded systems of mutuality and responsiveness, producing respectful adaptability as we explore conversational orientation in which to integrate our analyses. We can allow each other rest, and seed encouragement in each other to seek life-enhancing activity.The dual aspects of this double triquetral meeting in reality may be called its *physiological mindfulness*. Within physiological mindfulness, the presence of rational options creates the need for choice, and the potential for hesitation. The final integration of personal orientation to construct choice options has occurred in the motor-planning cortex, prior to the affective weighting of the final decision between them for core body action in the limbic core brain. If the three sensory association areas have been functioning in a fully yes-yes-yes triquetral balance, then the motor cortex receives a balanced informational input. The relational person making choices for action or rest needs to be 'present' at a heart level, however, with a mindful molecular and informenergy physiology that is poised for *timing*. The person explores balance in their movement as they choose the *timing for their decisions*.

The Zen image of the archer becomes relevant again. In the core brain, the arrow and the archer are informationally one. The tensioning of that informenergy pattern is its substantial understanding. The timing of a change of relatedness of this patterned form is a heart-level decision, one that integrates all three analytical perspectives in its movement. Choosing to cast out or release that inner tensioning of the prepared archer occurs at a critical point within the *timeframe* of feedback that builds the affective preparation state. The moment of chosen release has a *linear* purpose in a wider environment that converts affective preparation of a stative condition into effective bifurcation to an active state. The aim of that purpose gains meaning because it is set within the *eternal* perspective of that archer's personal substate fulfilled in that moment.

When deciding for intra-action in response to environmental change, all three perspectives on life can best be brought into a yes-yes-yes balance, with a willingness to move on with information loss to a renewed but as yet unknown emergence of life.

Balancing molecular and informenergy physiology

If the two feedback adaptive systems disconnect their parallel informenergy dynamics in the core brain, a wide range of mental states may follow, from intelligently over-analytic philosophical dualism, to dissociated psychotic or hallucinatory states, and all the varieties of human exploratory experience in between. In all of these mental states, the parallel systems may reset into healthy exchange by wilfully coming back to your senses and choosing action that reconnects your pulse-breathing patterns. In English there is a phrase, 'finding your feet again', that describes the grounding upon which people can balance to physically shake off unhelpful values and to disperse anxiety biochemistry in the present moment. This preparation for renewal allows the pace of change of the molecular physiology to reconnect with its underlying and perhaps racing informenergy physiology. Some people call this 'waiting for their soul to catch up with them'. It only needs a gap in time in which to find a safe place, at peace with the world, in which to rediscover that state of belonging together.

As previously described, at a molecular physiology level the core and limbic brain is closely associated anatomically and functionally with the brainstem, from where ascending activation of the cortex is co-regulated with pulse and breathing rates. This anatomical arrangement aligns the physical pumping heart and all autonomic inner functions of the body with the varying informenergy dynamics descending into the core brain's anticipations and ascending from there.

It may help to picture the *molecular physiology* to recall here the informenergy structure of a developing human embryo described in Part I. In adult life, informational flow will follow the same patterns of organisation, albeit in an anatomically developed body with a molecular physiology that is socially responsive in changing physical environments once the umbilical cord has been cut. The outer layer of the embryonic tube, the ectoderm, has a pattern-recognising capacity found in the somatic nervous system and the immune system of adults, which maps environmental changes. The inner layer of the tube, the endoderm, becomes the digestive tract, which develops its own autonomic nervous system. These two nervous systems have a fuzzy boundary with the embryonic middle layer, the mesoderm. The mesoderm has a different communication system. It is the blood circulation, into which the endocrine glands release hormones that feed into every cell of the body via the physical pumping heart. The physical pump heart has its own miniature internal nervous system, simple as that intracardiac coordination of cycles needs to be for a lifetime. Nevertheless, the cyclical coordination and timing in the physical heart is vital to life. It is highly responsive via hormones and autonomic nerves to the core limbic brain's coordinating activity. Life therefore progresses by this smooth integration of hormone chemistry, autonomic function regulation, and distance pattern recognition. All coordinates as the mesodermal shaping of the organism's movements, producing musculo-skeletal poise in preparation for activity or rest, which is embedded informationally within the wider ecologies with sensitivity. This is the *molecular physiology*.

An *informenergy physiology* is supporting all that molecular physiology, partly as biophysics moving among its structured forms, and more importantly as the wider *social sensitivities* that extend aware humanity beyond the fuzzy boundaries of its multiple relating bodies. The body chemistry thus belongs in the universe, not just to the individual. Beyond any dualistic understanding of local space and time or the substance of reality, a person's life is moved synergistically from within by what is beyond.

The comparative nature of consciousness can veil this informenergy physiology of belonging in the wider whole. The state of physiological mindfulness described earlier can unveil sensitivity to its movements. This deeper informenergy physiology may be intuited below the orientation horizon as a sensitive awareness of remote changing relatedness, which may inform the inner heart. In the orientating cortex, changing relatedness becomes converted into object recognition. However, that veil of objectivity disperses when making the wilful choice to love life, in its broadest sense of responsive connectedness into the harmonics of holomovement. That choice to change the focus of attention is an informational process that will synergically affect the probabilistic functioning of the whole body *from within*.

However, heart-level sensitivity becomes life-enhancing only if the enduring connection called love inclines our thoughts also towards intelligent, responsive kindness in relation to those who differ from us. Traumatising experiences can make people dissociate from the potential for kindness to contribute to their healing. They may seek safety by making artificial protections. On the other hand, if self-centredness prioritises gain by oppressing others, rather than gift, then 'what I want' distorts further the informenergy physiology. Such an attitude will counteract synergically radiant life, which others will detect subtly as a sensing of an emergent darkness in the informenergy physiology exchanged below their orientation horizons. Among people who know the substance of a stable domain in their inner hearts, providing the balanced presence of intelligently loving kindness to plan their responses, that sensitivity can be the source of a healing response, both to the traumatised and to oppressors. Stable domains are not platforms to rescue others, however. They are informationally substantial environments that are simply *relational*. Others can sense they are nearby, at hand, present at a place and a time where they will be welcomed to belong when they choose to respond to the call to relate from a stable domain of one who loves them.

Sensitivities reimagined enlighten consciousness

Below the orientation horizon, life is a bigger picture than any individual. Life-enhancing decisions when facing choices need to be focused on the informenergy physiology of synergic renewal, on the mutual light of awareness, not on the comforts of molecular physiology. Diversity and mutual respect, hearing each other's stories with sensitivity at a heart level, action-planning and resting in stable domain safety or security... all these features of life add to the relatedness quality that is sufficient for life-enhancement to restore synergic balance.

Episodes of shared awareness of emotive states may arise between people. They look at each other and know. Knowing states, at a heart level, may arise beyond the local space-time-reality ecology. Imagine that two physically separated zones of the cerebral cortex have a reverberating, bootlaced connection, establishing a harmonic informational cycle between them. A change from another source that intersects informationally with that harmonic could create an interference pattern with that cycling harmonic. The resulting pattern would informationally 'know' the quality of relatedness in that harmonic. This triquetral model does not exclude the possibility that the cortico-cortical neural feedback loops in the forebrain may thus create a visually 'moving picture' holographically of informenergy in the perceived reality. The physical environment of the brain in its bony skull, with its constant glucose food supply and constant temperature, is a highly stabilised ecology for informational patterning. It is potentially a unique triquetral phase space. There are exceptional states where informenergy thought patterns may be seen visually by others projected out from the cranium. Exceptional states also may harmonise remotely, giving unusual knowledge beyond the senses. Some people are more knowing than others.

A person's range of sensitivities determines how extensive or limited their local world is. Personal memories shaped by those sensitivities affect anticipations, both hurtfully and hopefully. Their resonance lights up the comparative patterning of conscious mind's thoughts initiated by that range of sensitivities. All of that bundling of informational activity from memories may come crashing down, however, when it meets the cascading input from the core brain's preverbal awareness of the way life really is impacting the inner heart.

Of course, people can resist the loss of informational patterning that may result, hardening their inner hearts to preserve preferred attractor states. This may be an honourable process when taking care to guard life-enhancing values while letting go of death-enhancing ones. Unfortunately, the opposite dishonourable process may also happen. Informenergy may build up around death-enhancing values to preserve the ego. Analysis takes over on how to avoid the emotions arising from reality, resulting in information loss from the social feedback by screening out sensitivities. The resulting dissociated analysis usually explores how to avoid love, compassion, or kindness, all of which are the informenergy of adaptability and responsive learning in wider ecologies. Coercive control, violence, and the corruption of healthy, nurturing order may follow, all of which further enhance grief emotions to perpetuate the death-enhancing cycle and dissociate from true heart-level values.

When emotions are understood instead as the evidence of potentially evolving personal values, the informenergy can be released from potential into actual movement to name them and shape life around them. Then the need for an informed choice in favour of life-enhancement becomes clearer. It happens when the useful purposes of unpleasant loss emotions become clearer. Responsiveness in a social ecology to mutually name personal values can restore meaning and direction for the timeliness of living together.

Being able to name life-enhancing personal values is a core life-skill, as important to learn in schools for social integration as literacy and numeracy. The capacity to do that naming is vastly magnified, however, in synergic conversations with people who see life differently, but who know also that there can be a healthy unity of shared adaptability in changing environments.

Part IV

Belonging Together in a Land of Quantum Gravity

For over 300 years, the Science of the Enlightenment has pushed educated people to be Observers of life. Now the New Science invites people to step back in, and enjoy sharing life with intelligent kindness. The three analytical perspectives on life add understanding to the dynamics of enduring love, which is our heart-level connection with open-system movement in the implicate order. This could strengthen the human ability to remain present with those who differ, but grief and anticipatory worry may drive people to separate into defensive isolation as closed systems.

Perhaps a higher-order understanding of how diversity in unity is creative may enable people to trust the implicate order enough to explore conversationally how to bring that implicate-order life into our shared explicate order.

Chapter 11

Movement Shapes a Quantum Gravity Universe

In his remarkably accessible book, *Reality Is Not What It Seems: The Journey to Quantum Gravity*, Carlo Rovelli communicates the historical path of science with such lucidity that his readers can easily identify at which stage in the journey they learned science when in school or college. They can hop onto the train he drives that takes readers to the limits of the current knowledge of the world as in 2017. Remaining uncertainties concern how the two fascinating extremes of scale of this shared world connect with each other. The cosmic universe is known to be expanding at an accelerating rate even now. In the micro micro-world, non-local quantum packages of change swarm synergistically and connect in unpredictable, seemingly non-deterministic ways, and yet they scale up to shape the basic building blocks of the consistent and predictable physical world that we share. This is David Bohm's transition from an implicate order enfolded at quantum or pre-quantum level unfolding into an explicate order where the rules of physics provide a structure to this implicate-order wholeness of movement. One tantalising question is left in the air, so to speak. It is how the thinking human being belongs in a physical universe that now seems to be swirling with information at every level between these far distant reaches.

I am not going to retell the narrative journey that Carlo Rovelli shares. It is there for you to read any time, and I encourage you to do so. Instead, I invite you to step out of the train at the currently final destination he describes, and to look around at the way the world seems now, recognising that our minds are inclined to think in ways guided by our inner heart values. Each one of us is a contributory part of this land we step into. Together we shape its life from within, even as we look out to explore and understand our macro and micro limits.

In Part III we have seen how the explicate order of conversational orientation in healthy human and physical systems can seem very different when people step back to analyse in three different perspectives. Enlightenment Science constructed a viewing platform above this land, not in it. From this elevated position, people could observe the land's forms and functions as if they were linear, predictable, controllable processes. Reacting against this Structuralist and Materialist view are the free thinkers—postmodernists who have deconstructed that viewing platform. Instead, they have taken to heart as individuals the mental distance it provided, imagining this to be their own inner space, their personal bubble from which they optionally share information to shape the movements of life. Alternative to both of these worldviews is the Third Way that updates ancient Vitalist views of an esoteric energy or spirit coursing through space and all souls as the true substance of life, which lies behind mere illusory mirages of a land that only seems material.

The sky on a clear night shows the land to be a richer and more mysterious context for shared living processes than is offered by each of these three analyses. Space is somehow deeper; time more eternal; the experience of living can include a substantial presence of peace at heart in the midst of its changing qualities of relatedness.

Part IV of this book illustrates how an informational view of the holomovement shared in an implicate order can provide a uniting substrate of synergy to account for mutuality, from which the explicate order of energy-mass emerges and diversifies into its localities. The many diagrams that follow show synergy prior to energy. They show a strong fabric of bidirectional communication—resonance in which harmonics can span distance and time to shape substance. This may be thought of as enduring love behind the potentially conscious explicate order, emergent as movement unfolds to diversify life without breaking its implicate foundational unity.

The extended illustration that follows, using networks of informational triquetrae, can show also how the explicate order can experience separation, brokenness, and misunderstanding without breaking the implicate order of wholeness. Healing and renewal then emerge as real possibilities, allowing the restoration of life after destructive change in the explicate order.

A quantum change

Sir Isaac Newton saw something strange working invisibly on objects that could accelerate their movement *as if from within*. The word 'gravity' comes from the Latin for weight, seriousness, heaviness. The universe is now known to be expanding at an accelerating rate, as if from within. We need to think gravity through again, as if part of an implicate order prior to other forces of nature. The triquetral model may have something helpful to give direction to those thoughts and the deeper values that move them.

Whether or not Sir Isaac was actually sitting in a chair, watching ripe apples fall spontaneously from trees to be stopped with a thud by the ground as he shaped his mathematical ideas of acceleration and momentum, is irrelevant to this next fact... Presenting scientific dilemmas and solutions in such a story adds colour, movement, sound, and the possibility of taste to an otherwise potentially barren and disjointed landscape for many readers. The narrative of science is storytelling in the language of mathematics, which excites those who can think and communicate in that language, but it leaves others out in the cold—except that in Chapter 1 we initiated a thaw by noting how the same triune principles of organisation can be seen in mathematical equations as in every other specialist and even artistic language.

The New Science of renewed order emergent from transient chaos has gone well beyond the classical mechanical view of the universe that Sir Isaac contributed to developing. The mystery of how objects that seem to be solid to our senses can have spontaneous inner dynamics—including pulls that act also on inner pips and bruises and maggots—has been progressively solved. The New Science brings together the cosmic-level picture of gravity holding the whole universe together in a rotating dance, with the ever smaller understanding that apples are made of cells that are made of molecules that are made of atoms that are made of particles that are made of electromagnetic fields in which waves spark *intra-active change events* that, to an observing scientist's micro-camera, look like the realities we call 'wave-particles'. More generally, these wave-particles have been erroneously called 'quantum packages of energy'. The triquetral model of quantum gravity will call them 'quantum changes'.

Your mind may prefer the solid concept of a particle, or the dynamic one of an energy wave, but the triquetral cosmology model gets inside both particles and waves to see them equally as diverse patterns of synergy structuring the changing relatedness of movement. The triquetral model clears a way to remove positivist images of packages of energy from our thoughts, and to replace them with process images of intra-active relational movement, in the synergies of which we participate.

Scaling down to an informenergy view of nature

We started at the mid-scale of personhood to describe in Part I an informational view of life's experiences. This is evidenced in neuroscience, in learning theory, and in adaptive communicating systems, where principles of emergent order by cycles of feedback lead to resonance. We looked at how informational feedback loops in the brain orientate an aware person. Below the orientation horizon of the conscious social ecology, the human being is connected into the continuous movements of the wider physical environment in various ways, with more or less sensitivity, awareness, and intuition. Here we shall explicitly connect the human inner heart with the implicate order of wholeness.

As an example of the sensitivity of where the inner ecology responds directly in the wider physical ecology, consider the retinae of Sir Isaac Newton's eyes as he sat and observed the apple tree. The rods and cones of his retinae would electrochemically discharge when a protein molecule there, called opsin, was triggered to transform its shape on relationally receiving even a single photon of light into its structure. Millions of photons would be transferred in watching apples fall to the ground. They were organised into patterned forms across his retinae, which relationally transformed as the apple moved. These changes of relationship of patterned forms would have been converted informationally, via his optic nerves and various ganglia and visual cortex, then through the sensory association area of his temporal lobe, mixing with other sensory associations in his motor- or action-planning frontal cortex and his core brain for balancing, into the notion of a self-existent apple as an object that incorporated all his past experiences of watching and handling all apples. Without knowing it, he would have been in a mindful, bidirectional informational exchange with that apple as it fell, in which his presence contributed to the apple's unknown environment. This background resonance to the observed fall could be represented in a double triquetra.

That iconic representation of his presence has four side ligands. We have previously noted that these ligands are the informational means by which unpredictable influences from diverse sources can influence the ongoing synergic process at any scale of life. These remote influences may emerge into the explicate order as an activation of Sir Isaac's power of choice over his potential response to watching the apple fall. He may consider picking the apple up; noticing how solid it seemed to be, from whichever angle he chooses to look at it; he may consider rubbing it. He may have considered its weight in his hand; then decided to bite into it, pleased to find there was no maggot in the crisp new surface remaining. The apple had no capacity of choice. Sir Isaac did. Deeper than that diversity, however, is how the observational Sir Isaac and the apple were already in a shared synergic bidirectional informational flow, which in the implicate order moved Sir Isaac to potentially pick up and repattern the apple while he mindfully wondered how the wider physical environment could accelerate it down to the ground from the tree. Were birds singing and flying from branch to branch, defying the gravity that he was beginning to imagine? Was there a wind or a breeze blowing in the autumn sunset that pressed small leafy branches to move and flutter as he tried to make sense of the mismatches of his observations with his ideas and anticipations? The first evening star might appear, and the moon be seen already high in the sky as he heard another apple fall to the ground.

How the power of choice scales triquetrally in energy-mass

Every sensation experienced by Sir Isaac Newton as he sat or considered that evening is an informational exchange that scales down to quantum changes in the implicate order. The triquetral model asserts that these quantum changes do not require an externally applied force to initiate and differentiate in spinfoam movements. This is because the foundational relatedness principle means that synergic movement is intrinsically *within* the waves and particles of materially structured life, not caused by them. They are manifestations of the synergic mutuality of movement between both 'ends' of communication through a spinfoam of wholeness. The diversity that leads to choice options comes through the side ligands of a double triquetral mutual change. That is the nature of gravity too, as will be further explained using the triquetral network models. Gravity does not need to be imagined as a *force* of acceleration or attraction acting on a thing. Gravity may be a network of movement, not even the curvature of space-time, that holds together and introduces diversifying changes in different environments, both physical and (as we shall go on to see) mental. Through that synergic spinfoam network, forces might be imagined as higher-order movements. These movements and events emerge in measuring instruments as changes that scientists have called wave-particles.

The relational flow informationally through this proposed triquetral spinfoam is bidirectional. It is a synergic reality through which linear, unidirectional higher orders of movement may optionally pass. The implicate order understood in this way is one of a resonant feedback network everywhere, a spinfoam, extending that resonance to its utter limits before the emergence of space, time, and substance. Through this synergy of life as depth movement, holomovement, additional dimensions and scales of energy waves may progress. Progressive waves may seem to be unidirectional in space-time-substance, and therefore may be compared with each other and therefore potentially measured. But that unidirectionality is only a timeframed transition before those higher-order changes reflect back through the synergic spinfoam. Every action has an equal and opposite reaction. It is just that higher-order progressive movements are slowed through the synchronicity of a resonant network by all the triquetral side ligand effects. This limits the speed of light in a 'vacuum', in a spinfoam. The spinfoam resonance of movement is incomparable, however. The wholeness out of which timing has emerged, mingling with space and substance, is not measurable. Wholeness is/does extensive fullness and peace.

When a measuring device is introduced from the explicate order into a locality of the implicate-order spinfoam, a movement in which the experimenter is consciously active and making choices of behaviour, the spinfoam bidirectionality will now include the instrument's internal patterning. This inclusion in bidirectional presence of wholeness would allow some linear energy input to flow. The measuring instrument will therefore alter the patterning of linear movement emerging from the spinfoam, as we shall see later *prior to* its measurement as a wave or particle. The bidirectional resonance of spinfoam will, nevertheless, remain unbroken by this energy flow as the wholeness from which any linearity can emerge.

Now, a person choosing to insert the measuring instrument into that locality of spinfoam is, by definition, a substate of that wholeness. At a heart level, their personal values are a non-local connection with all holomovement. As previously described, personal values *are* feedback cycles of moving informenergy. They are structured in neural network filters in the core limbic brain and transform whole-body physiology through neuro-endocrine chemistry and physics. These values emerge in the social and physical ecologies as action and behaviour. A personal value is therefore somewhat like a measuring instrument introduced into life. It extends from the explicate-order clash of inner ecology feedback loops with social ecology feedback loops, the dynamics of which become present in the implicate order's pre-existent synergy. Personal values and their associated choices influence the same spinfoam resonance in which the person's physical body and physiology participate, whether handling a measuring instrument or discussing action plans with colleagues. However, that person, when focusing attention to plan, becomes to some extent insensitive to the wider wholeness of the implicate order. Their inner repatterning embeds different heart-level values while planning an experiment *on* wholeness to measure its linear emergence. When that person takes a mental step back to analyse the experimental measurement, they close their mental system away from the open system of their inner hearts. They become part of the local connection, part of a closed system that includes their measuring instrument and the linearity that is moving through implicate-order wholeness, like the archer with the bow and its arrow.

If people do not love life, and do not choose to love valuing life in all its diversifying capacity, and do not choose to heal rifts and divisions and hatreds that are set up competitively within the explicate order, then thought cannot compensate or bridge that chosen separation from life into a world of analytical perspectives. Thought separated from heart-level values creates closed systems of ideas. Thought alone may break wholeness.

If, however, the inner heart with its embedded values is orientated towards choosing life, then thoughts that partner with that heart-level connection into implicate-order holomovement can transform explicate-order repatterning. A more cooperative movement may emerge, opening a way for greater curiosity about other people's values. It is a matter of choice. The resonance, potential harmony even, of that bidirectional open-system feedback of a person with their environment could transform personal values, and their associated material transformations of the world, if an occasional gap in time is allowed to influence closed-system thoughts in a more open, life-enhancing direction.

There is a balancing paradox here that affects the frame of mind in which people make their choices. Linear energetics running through a spinfoam of resonant quantum changes do scale up and emerge, or unfold, into an informationally predictable, deterministic world. We shall go on to see more clearly how that can happen. Despite the Aztecs' sad reasoning from their collective explicate belief and ideas about cause and effect, we do *not* need to encourage the sun to rise each morning by making human sacrifices. Fear shaped many people's lives and deaths in that culture, because they misconstrued how their thoughts engaged with the explicate-order world. Cycles of material-energy processing *are* independent of our local presence and action at an explicate-order level, even though at an implicate-order level, within these processes, our heart-level values are dynamically embedded non-locally. Locally, human choices do, however, influence the physical and social ecologies, especially when they have been framed in a closed-system mind. So, the relevant personal growth question is this. How does an individual person constructively balance within their inner ecology the 'power of choice' benefits of analytical reasoning locally with a non-local conversational perspective on life-enhancement? Locally made personal choices will spread their influence resonantly into the wider non-local features of the physical and social ecologies.

The answer is to include activation of Emotional Logic when creating an occasional gap in time as a safe place in which to reflect thoughtfully, then using a yes-and-no-no analysis of imagined impacts on inner peace of adopting different perspectives on life. In Chapter 12 this alternative to mind-emptying will be explored more deeply. At this stage, the purpose of mentioning this strategy is to rehumanise the way people think about the implicate order. This level of orderly movement is not a mere physics of forces or numbers. Implicate-order movement *includes* personal values shifting between their joy and grief modes of love as people connect or disconnect their lives. In the implicate order, the dynamic nature of personal values *feeds back* into what matters. The implicate order has many substates, or localities, some of which like human beings have the power of choice over how to relate.

3D-printed models of triquetral informenergy

Reasoning thus, from personal neuroscience in a social ecology to the quantum science of a physical ecology, I started to make 3D models of triquetrae to see how these processes might cohere.

I found that pipe cleaners were too flexible to sustain a consistent form to represent the triquetral process. When I tried warping wire instead, I realised how much subtlety was informationally going on in the warp and weave of three semicircular loops as they create the 4D phase space of the triquetra's central zone. I decided to move on to 3D printing to explore how two triquetrae could connect with each other, and into more extended networks. Fortunately, just 20 minutes down the road from where I live I discovered a commercial company, Tooltech Ltd, that made 3D-printed components for British Aerospace. It turned out that I knew the owner. It was a family business, so his son and wife were there also, organising their dozen employees and trainees. Yes, they would be delighted to help me explore how this new quantum cosmology theory might be modelled. They too were curious to see how it might shape the world informationally from within particles.

Andrew Newcombe instructed his trainee Ricardo Pereira to do the Computer Assisted Design (CAD) using the advanced Siemens NX software. They later said that these were the most complex models they had ever printed. Cleaning off the support materials to reveal the detailed loops and sine waves gave a relaxing break from the hectic workload that otherwise kept them busy. Ricardo worked hard to understand the model, working out the algebraic difference between a triquetra (semicircles connected unendingly at angled points) and a trefoil (an unendingly coiled ribbon). He went on to shape its relational substance, first of all digitally, then trimmed to fit in the printer. We opted for white plastic as the material, which I would subsequently paint with acrylics to show the colour coding of *informational movement* at any scale. The *changing relatedness of patterned forms* is consistently shown as blue for change, yellow for relatedness, and red for patterned form. In the figures reproduced here, these are shown as dark, light, and mid-tones of greyscale, but as mentioned in the List of Figures there is a colour PDF that can be downloaded for those who would like to get to know the model in depth. We are about to discover how practically to *avoid* seeing these principles of movement as objects in themselves.

These models all show orderly movement. They could be made of rubber to show how every feature could be vibrating, wobbling, or spinning as qualities of relational connection shape patterned movements of the explicate order informationally from within. This is a philosophical statement about process ontology. *Movement is innately ordered*—not a primary chaos. Mutual transformations within the resulting informational spinfoam arise from a sacred set of simple rules of adaptability, from which life movements emerge. Our temporal lobe sensory association areas turn these relational changes of movement into objects. This enables people to manipulate them, and to talk about them in a social ecology, but in so doing our cerebral cortex may hide implicate-order reality.

We watched patterns of interconnecting movement extend across the screen, in front of our eyes. Triquetral understanding entangled there into patterns of movement that adopted unexpected forms, some of which could be the seeds of substantial particles, and others the progressive energy of waves connecting them. On a digital screen we watched analogue change moving patterns of resonant vibrations. These vibrational movements could shape the world from within. It was an awesome experience.

***

The core icon is the double triquetra, shown in Figure 5 in Chapter 2 and replicated for comparative purposes in Chapter 9. One triquetral change is an ecology for another in unending movement. One informs the other of its change from a neighbouring location. That relational connection is, in applying this icon to the physical ecology, ordered movement below and beyond our human mentally constructed orientation horizon of space, time, and substantial experience.

The four outwardly directed corners of this iconic vibration are potential ligands having different qualitative effects in an extending spinfoam of entangling triquetrae. If the relatedness ligand shown connecting the two changes were itself to transform or break when remote changes entangle with those corner ligands, then change will occur synchronously at both ends of this resonant double triquetra. If observed from another scaled-up location in an explicate order, these synchronous changes will seem to be remotely unconnected in space and/or time. The resonant double triquetra, we may presume from within an explicate-order mindframe, can seem stretched in ways we cannot yet imagine above our space-time-substance explicate horizon of orientation. Each change in this double triquetra will have further consequences that informationally move on into the local environment of events via extending implicate ligand chains.

The patterning of these remote quantal changes may make little sense when considered by scientists in Oberver mode above our orientation horizon. For example, a change of relatedness *event* seen on a Hadron Collider photographic plate (after remote acceleration of quantum changes under the mountains) may be called a particle. This enables a scientist to describe its *relational frame* as its potential behavioural properties, which becomes summarised in its name. An Observer may use that name to define 'an object'. However, that 'particular change event' will have had a remote analogue source, namely the whole scientific establishment excavated under the Alps by the intentional choices of scientists, engineers, and politicians. That whole establishment could be represented by a single triquetra that represents the neighbourhood ecology within which a scientist (another single triquetra of self-organisation) observes the collision. The locality of the photographed change event, the collision, had been *informed* by a coordinated diversification of localities elsewhere. Such is the transforming nature of a relationally changing analogue cosmos.

The first 3D models we made were of a double triquetra, a six-triquetra stable domain, and a helix derived by stringing triquetrae together. They looked interesting, but something was wrong with them. As I looked closely from every angle, I noticed that the ligands did not align in both directions. This was most clearly visible in the stable domain. One of the loops made a smooth sine-wave flow into the next triquetra, but the other loop kinked at each junction. This could not result in resonant, bidirectional flow of informenergy. We thus discovered that the original 2D images of triquetrae had missed an important feature, which opened the way to a most astonishing new understanding of substance.

Figure 7: An l-spin triquetra

Figure 8: A d-spin triquetra

On researching 'knot theory', I found images of the triquetral knot in its two well-known patterned forms, a clockwise spin and an anticlockwise spin, as shown in Figures 7 and 8 and replicated here. To recognise them, take any corner of a triquetra and follow the loop down to the right. If it passes behind the crossing loop, it is a clockwise l-spin triquetra. If it passes in front of the crossing loop, it is an anticlockwise d-spin triquetra. Amazingly, this core diversity turns out to be the informational source of the differences between waves and particles emergent from the implicate informenergy order.

Three types of double triquetra result from different pairings of l-spin and d-spin triquetrae, as shown in Figure 24. In the centre of that picture, an l-spin and a d-spin triquetra entangle, so the double triquetra includes *diversity*. The bidirectional central ligand then holds all four corner ligands in a 2D horizontal plane, except that this plane has a thickness that depends on the curving height of the internal sine waves that criss-cross the inner vibrational movements.

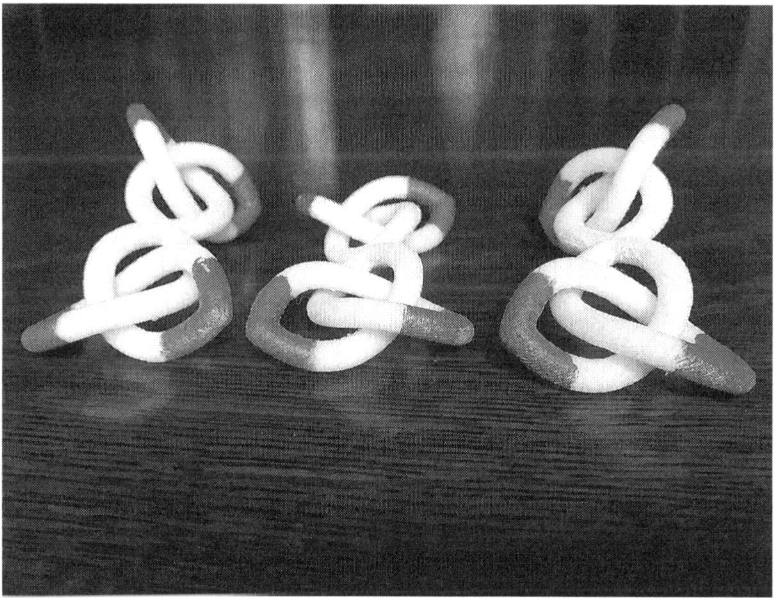

Figure 24: l-l-spin, l-d-spin, and d-d-spin double triquetrae

When two *similar spin* triquetrae entangle, however, the connecting ligand has to twist to establish the same bidirectional informational flow. An l-l- double triquetra twists the four corner ligands clockwise (seen in Figure 24 on the left). A d-d-double triquetra twists the four corner ligands anticlockwise (seen in Figure 24 on the right). These rotational twists hold the four offered ligands in different orientations.

In this we see the origin of the three dimensions of space emerging out of vibrational movement at a quantum informational-change level.

All three double triquetra patterns have equally resonant states of vibration. They could scale up through the corner ligands by further entanglements of quantum change into diverse extending patterns in the implicate order. These extending patterns could result in a spinfoam that has three different transmissible states of harmonic vibration running through the holomovement of informational connectedness, as we shall go on to see.

Quantum computing requires three states of quantum change. These three triquetral informational patterns resonating at the Planck length could possibly correspond to the +, 0, and – qubits. However, these are not 'digits of information'. They are not self-existent nouns. They are mutual informational changes that provide a network of synergic connectedness through which higher-order movements could transmit.

I shall call these three states of a double triquetra *informational qubits*. They are not uniform, regular objects or spaces that move only in relation to each other. They are analogue informational vibrations with stretchy ligands beyond our concepts of space and time that can snap and remake in different patterns. They are *not* 'qubits *of* information'. Informational qubits are relational quantum changes in the holomovement synergy of implicate wholeness.

Space

A double triquetra has four potential ligands that can further entangle in 3D space. If double triquetrae with similar twists (+ or – informational qubits) gather from an environment and entangle into an extending triquetral synergic standing wave (or a string), then a wave of progressive movement could additionally extend along the string, like a slinky. That progressive wave could represent the conversion of synergy (stative waves connecting a spinfoam) into progressive energy. Energy could thus have additional remote effects in localities of the implicate-order spinfoam, which effects are informational—informenergy. That progressive wave would be a fourth dimension of movement within a 3D spinfoam of harmonic wholeness. When scaled up, these fourth-dimension waves would be knowable as space-time. Progressive energy waves are therefore, in this triquetral cosmology, a different dimension of order from the triquetral synergic standing waves of the spinfoam. Triquetral waves are synergistically linked qubits, which form a 3D substrate for higher-dimensional types of movement.

Figure 25 shows a string of d-d-qubits made when *relatedness* (yellow; pale) and *form* (red; mid-grey) ligands alternate. A helix of synergic, bidirectional movement between its two ends results. This iconically represents a triquetral standing wave.

Figure 25: A d-twist triquetral standing wave of synergy

Tracing carefully along the yellow-orange-red-orange-yellow string in Figure 25 etches a complex, revolving, helicoid sine wave of movement into the universe. The *relatedness of forms* (yellow-red) is the structure of *space* philosophically. Gravitational waves are predicted by Einstein's General Theory of Relativity. They were first measured in 2015 from the collision (a long time ago) of two black holes. Gravity is no longer thought of as a force of attraction pulling two objects together, like an apple and the earth. It has been described instead as the curvature of space around massive objects, which influences movement. But that description ascribes a reality to space itself, which could sound a bit Materialist-Vitalist in perspective, giving me a no-and-yes-yes hesitation about that description.

To balance it as a fully yes-yes-yes triquetral picture, imagine strings of triquetral quantum changes in standing waves extending across the universe in a spinfoam network. This implicate order of synergic informational movement may provide a picture inside how that curvature comes about. In triquetral informenergy and synergy terms, a vacuum is not empty; a vacuum is alive with harmonic peaceful movement. That synergic presence of life slows down or changes tone into time as it patterns into mass and other physical features of explicate order. Activating progressive energy movement through this synergic spinfoam is an explicate order running through the implicate shaping of space 'around massive objects', although the massive objects have themselves also emerged from that synergic presence.

This informational model of the universe shows how the land is shaped locally to where we live on whichever upside-down side of the globe, and how apples fall from trees with a thud as they are stopped by the ground. Figure 25 may show the inner flexible structure of those triquetral gravitational waves. The connection between apple and earth and Sir Isaac's retinae is an extended network of these triquetral waves. Each triquetral gravitational wave offers blue ligands (dark in greyscale) outwards into the extending movements within the various localities that they extend through. Next comes a truly remarkable feature of this triquetral image of movement, which was discovered by making 3D-printed models. Blue represents the principle of change. A triquetral gravitational wave will therefore cause change *from within* any other movements that entangle with this wave, anywhere, within any diverse locality, synchronously, and substantially, stretching across the whole universe.

The hypothesis proposed is that triquetral gravitational standing waves are only scientifically measurable because of the change effects induced in a neighbouring locality as a higher-order energy wave moves additionally along Figure 25. Change in environments flows through that spinfoam network of stretchy ligands, enabling the gravitational waves to be measurable. This is the multidimensional nature of scaled-up movement. The notion of space and locality extends from the micro to the macro in which each of these scales has its patterning movement, all fractally scaling up because they all follow logically from the same simple triquetral rules or principles of organisation. Triquetral waves bring mutual, synergic, informational connection between remote localities at every scale.

In this triquetral model, there are two other types of triquetral wave that will be shown to correspond to electromagnetic waves and thermodynamic waves. Each type has an l-twist and a d-twist version. Astonishingly, these two types of informational wave scale up philosophically to construct substance and time, as we shall see, the two other primary neuropsychology constructs for orientation alongside space.

Substance

Figure 25 shows a d-spin triquetral wave with alternating form (red) and relatedness (yellow) ligands, and all projecting ligands offered being change (blue). Rather than show greyscale images of the two other corresponding types of triquetral wave, I shall explain their differences and similarities. However, in the associated downloadable PDF of colour images there are a (gravitational), b (electromagnetic light), and c (thermodynamic) versions of Figure 25.

When a triquetral wave is strung together with alternating change (blue) and relatedness (yellow) ligands, our temporal lobe sensory association area working on that notion would convert its changing relatedness into an impression of substance (or the experiential understanding of a real presence). Changing relatedness shown in this triquetral wave offers only red (form) side ligands into the spinfoam network of movements. These side ligands will bring a 'sticky' quality of patterning into the connectedness of the spinfoam. Entanglements with them will introduce the movement principle of *form* and qualities of coherence into patterns. Although the type of triquetral wave shown in Figure 25 looks as if it might be 'progressive', as if it might represent a flow of energy, in the triquetral model it is bidirectional synergy, an extended ligand like a standing wave of vibration creating remote synchronicity. It is an image of informational connection. The changing relatedness type of triquetral standing wave offers through its side ligands a strong force of attraction or repulsion that diversifies moving forms and gives them substantial connection in their movements. One quantal local ecology will therefore move informationally in a way that is responsive to another locality's remote movements.

All triquetral waves have l-twist and d-twist opposites. We could hypothesise that, when scaled up into our objectively knowable explicate-order world, these variants of changing relatedness may be the source of the diverse electric charge (+ve/−ve) and magnetic polarisation (north/south) that we measure as electromagnetic (EM) forces. With changing relatedness we are therefore probably looking at electromagnetic triquetral waves extending through the cosmos. With gravitational waves, this same diversity may contribute to dark matter. Opposing EM forces resist the collapse of the universe under gravitational influence. It is therefore this feature of triquetral cosmology that would give deterministic substance to patterned forms in movement. Objects, on scaling up synergistically, start to behave deterministically even though they emerge from an indeterminate (but not chaotic) quantal movement substrate. EM forces balance and potentially stabilise this rotating and vibrating universe.

Just as gravitational waves will have measurable helicoid movements along their length scaling up into several dimensions simultaneously, so EM waves will have measurable frequencies of longitudinal vibration. Some of these are in our human sensitivity range as visible light, because they activate rod or cone cell membranes electrochemically. They may even occur as pulses of activation—photons. Most of them are outside our human sensitivity range, and may even represent the strong forces that hold the atomic nucleus in its patterned form. This will become clearer when we start looking at how the triquetral cosmology model sees inside particles as well as inside waves.

Stars eventually burn out their hydrogen fuel and cease shining light as they collapse into intensely strong gravitational fields called black holes. These trap EM light within their shrinking event horizon as gravitationally they suck in distant planets and moons and grow in their dark intensity. They also belch out clouds of superheated particles, so the current theories go, telling the unpleasant stories of heavenly dramas unseeable in the clear night sky. Behind and within all these stories may be the informational feature of triquetral EM waves, now visible in their coloured variant of Figure 25. These offer their red 'form principle' ligands along their radiant way. This sort of triquetral wave introduces patterned form informationally into the shaping of cosmological movements. The attraction and repulsion of, presumably, l-twist and d-twist triquetral EM wave types both arise from the same synergic bond, however, forming spreading patterns among remote events.

Time and warmth

The third construct for orientation can be reconciled with the third type of synergic informational triquetral wave. Here, change (blue) and form (red) principles of movement are the alternating ligands. The *changing* of *forms*, especially cyclically, is the construct for the conversational orientation known as *time*. Cyclically vibrating patterned forms (objects) are also the nature of warmth or heat. The warmth or heat of a substance is its short-range internal vibration, both within forms and between neighbouring forms. This short-range vibration can increase in intensity to produce a change of state of substances—solid, liquid, gaseous. This type of triquetral wave is therefore probably a thermodynamic wave introducing changes of form into patterned substances. Instruments that very accurately measure time, such as atomic caesium clocks, have to be kept very cold, at temperatures where their degree of vibration can be constantly used as a standard against which to compare other vibrational changes. The same is true of quantum computers, where an artificial cold constancy is created in a local environment against which to inform the wider ecologies of the implicate-order movements that can inform everyday human living.

As previously described, patterned forms in a triquetral cosmology are internally coherent groupings of movement, held additionally by their side ligands with strong EM forces internally and externally that can resist gravity by acting over short distances. Thermodynamic waves would also run through that same spinfoam of quantum changes. Philosophically, these thermodynamic waves offer 'yellow' side ligands into the wider spinfoam ecology, introducing the principle of *relatedness* into its movements. Patterning forms will therefore also have this principle of synergic movement spreading through their internal patterning structure. Forms will thus have internal vibrational changes. By external relatedness, forms could become synergically coherent in their changes with neighbouring forms. For each form, its internal informational coherence will potentially allow an overflow of synergic movement out, to influence the dynamic patterning of neighbouring forms. This brings relational coherence amid diversity. In a closed energetic system, this feature would appear as entropy as it cools. In an open system, this scaling up of vibrational coordination could create the knowable warmth or coldness of coherently patterned and diversifying objects in our everyday world. These qualities of warming or cooling would emerge in the explicate order of our experience alongside other 'state qualities' such as rigidity or softness, durability or fragility, visibility or darkness, tastiness or sourness.

The scale of thermodynamic vibrations and spins that shape objects from within leads over time to changes of their substantial state. The most obvious example of a change of state is the way water varies between ice (solid), fluid water (liquid) at a range of temperatures, and steam (gaseous). In this triquetral cosmology model, these thermodynamic triquetral waves in the quantum spinfoam will be operating over the relatively short distances (in our human everyday terms) between the whole patterns of change called atoms, molecules, and macromolecular structures, such as cell membranes and cellular microfilaments (which are the internal 'skeleton and muscles' that move living cells from within). Relational movements of these changing patterned forms bring the feature of timing into the freezing or melting processes of changing states of the substantial world, with descriptions of timing using the three analytical types of time—linear, timeframed, and the eternal now.

Metals, as another example of states of matter, may be solid in our everyday experience, but they can be heated to the point where they turn liquid and then vaporise into gases. Heated iron is well known to glow with light. Simultaneously, it loses its magnetic polarisation and electrical charge in that change of state. Therefore, we know in the everyday world of our experience that triquetral thermodynamic waves (heat) and EM waves (light) interact. In this model the interaction occurs also with gravitational waves. These interactions will all occur by sharing the side ligands offered from synergic triquetral waves, which will entangle with other triquetral waves as they integrate into a shared spinfoam state of *implicate movement*.

Gravity meets light in triquetral cosmology

Triquetral waves are synergic connections into *mutual* movement. This is an informational process, relationally prior to the emergence of substance. Triquetral waves are not energy flow. They are the relational synergy that allows energy flow as a fourth dimension. Their side-ligand interactions may therefore explain the known influence of gravity on light. At the time of writing, no other cosmological model can adequately explain this interaction. This may be because the models do not make a clear enough distinction between synergy and energy.

Figure 26 shows the CAD image of a 3D-printed model that demonstrates how a triquetral gravitational wave (blue side ligands; dark in greyscale) and a triquetral light wave (red side ligands; mid-tone in greyscale) share a common connecting wave ligand of movement. That ligand happens to be the principle of relatedness (yellow; pale in greyscale).

Each wave *seems* to pass through the other. But here we must pause for thought. Most people will think in everyday terms of waves being 'progressive'. Waves on the surface of water are blown by the wind, externally pushing water's substance in ways that may eventually break up into a turbulence of spray if intense enough. It is a linear view of waves that in everyday explicate-order experience can be transferred to thinking about light and radio waves, which have transmitters and receivers. But triquetral waves are a level or two deeper than water and wind, and one level deeper than light and radio. The three types of triquetral implicate-order standing wave in this model—gravity, light, and heat—each have l-twist and d-twist forms. All are bidirectional *synergies* in informational movement that connect *synchronously* across *localities* of the cosmos, and unite them. The way they pull on each other sideways within the spinfoam network creates another order of interactive *energy waves* progressing along their lengths.

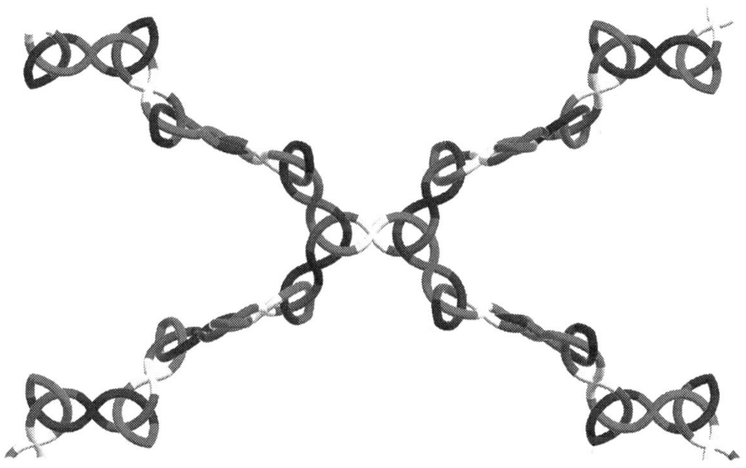

Figure 26: Gravitational and light waves share a relatedness ligand

These fourth-dimension energy waves running along triquetral waves criss-crossing the spinfoam become measurable when a structurally minded Observer compares their movements with some arbitrary standard. When an Individualist creates their own internal patterns among progressive energy waves, however, that person becomes an intra-active Creator of movement in a way that radiates information (noun) or informational change of patterned forms out to influence other localities. When a

Participator in life experiences an intuitive pressure to conform their behaviour to remote changes, sensed informationally within the spinfoam, these fourth-dimension energy waves may come to seem authoritative, rendering that person a passive recipient of order. But all three of these analytical approaches to energy waves miss the synergic depth of inner heart. The human person is already bidirectionally and mutually connecting in the informational patterning of those triquetral waves. This heart-level depth is intuited by artists and humanitarians, some of whom are also leaders in the sciences. They sense the moving connectedness of *life* behind the explicate-order 'it' of it all. There is an I-thou, or a me-and-you, to the implicate order of life.

The analytical perspectives are *not wrong*. Each has the potential to view the whole of life, but cannot give a complete picture of life. They are just partial understandings, perhaps in some way relatively heartless, missing the full depth of humanity in the physical cosmos. Figure 26 shows a single extracted feature of a densely vibrating network—a spinfoam of triquetral *quantum changes*. This is not merely microphysics. The changes are multidimensional core movements that scale up to enable all three primary perspectives to be consciously constructed in a mindfully human brain. M-theory would suggest there are 11 types of scaled *movement* making the hyperdimensional context in which conscious life emerges. Figure 27 is a photograph showing the full set of implicate-order models that cooperating, orientated, conscious, conversing minds sitting at a computer could discover and print. They show a range of patterns of informational movement that delighted our hearts and led to times of celebration. CAD images from this design process follow to explain how informational movement can relationally scale up into the shaping and reshaping of the energy and mass of explicate-order wave-particle substances, such as those that we imbibed having chinked tea mugs together.

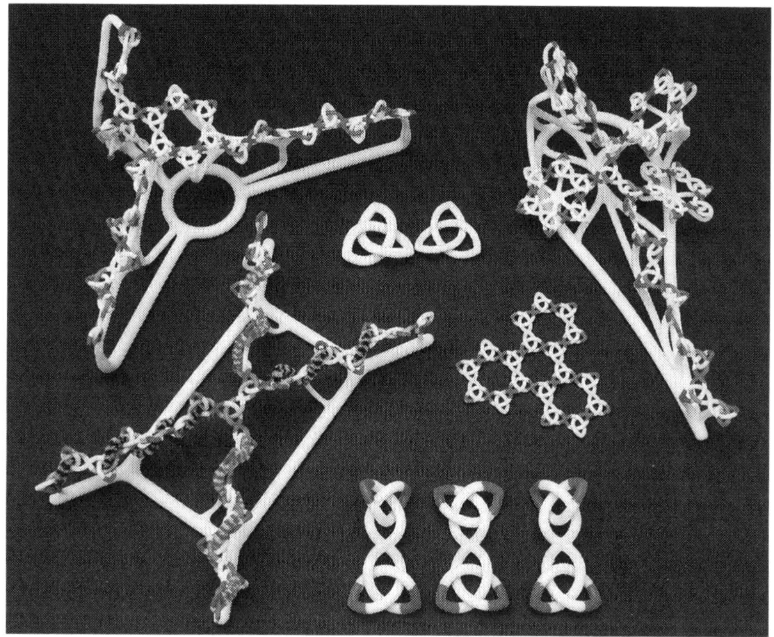

Figure 27: 3D-printed models of triquetral informational movement

It is important to accept that *quanta* in this network are not uniform. Quantal changes are relational and diversifying. They create spaces, timings, and diversifying substances that stretch and transform in their relationality. A vital conceptual step on is to accept also that such an informational network will have harmonic resonances that appear synchronously at both ends of the informational connection between remotely 'mutual source and receiver'. Any comparative measurement of changes will detect one extra dimensional step on from this meaningful level of triquetral synchronicity that makes an informational cosmos. Devices measure only the energy waves that are proposed to progress through this *synergic process network*. Progressive energy waves may create a change event remotely, which to a mind primed analytically to see it so may *look like a particle*. The event, however, is a remote synergic transformational change in a stretchy quantum spinfoam of triquetral waves.

Energy waves moving through the triquetral synergic spinfoam would be slowed along the connecting ways by the nature of their side interactions. The gravitational network could slow light, and warp its path as if through a gravitational lens placed between the remotely synergic localities, such as massive galaxies. It would do the same to heat, as it does in stars. In stars such as our local sun, gravity is pulled inside the fusion of hydrogen into helium atoms. That transformation of patterning releases spare heat and light. Internally to the forming helium molecule, the resulting local mass slows the progress of energy waves through the triquetral spinfoam of a star, creating strong forces within the atomic nucleus; but externally, those same fourth-dimension EM waves transmit through the spinfoam 'vacuum of mass' (called empty space-time) out to the far reaches of the cosmos, slowed and warping as they encounter planets and galaxies far beyond. A triquetral synergy view of the networked cosmos may explain what is currently understood as the speed of light. A synergic network picture of a vacuum may fit better with the special relativity of light's speed than imagining a linear force of packaged wave-particles travelling through nothing.

***

We would do well now to fill out the notion of substance in this view of the sun warming our local solar system's remote lands. How do planets and moons with physical mass, and particles that decay, emerge in this triquetrally synergic view of the cosmos? Particles and mass are not imaginary, even though a Vitalist might conclude that they seem to be. They are realities that significantly influence our powers of choice from within it all.

A range of particles emerge from extended SLMs

So far we have been looking at entanglement patterns of l-l-spin or d-d-spin triquetral waves (+ and – qubits). Something different happens when diverse l-d-spin double triquetrae (0 qubits) entangle with more of the same diverse kind. They remain in a 2D plane, not in a helicoid string. They form a dynamic hexagonal pattern—a stative movement with a swirling informational thickness that depends on the depth of inner curve of their movement loops. They form a stable domain, an SLM for short—**S**tab**L**e do**M**ain.

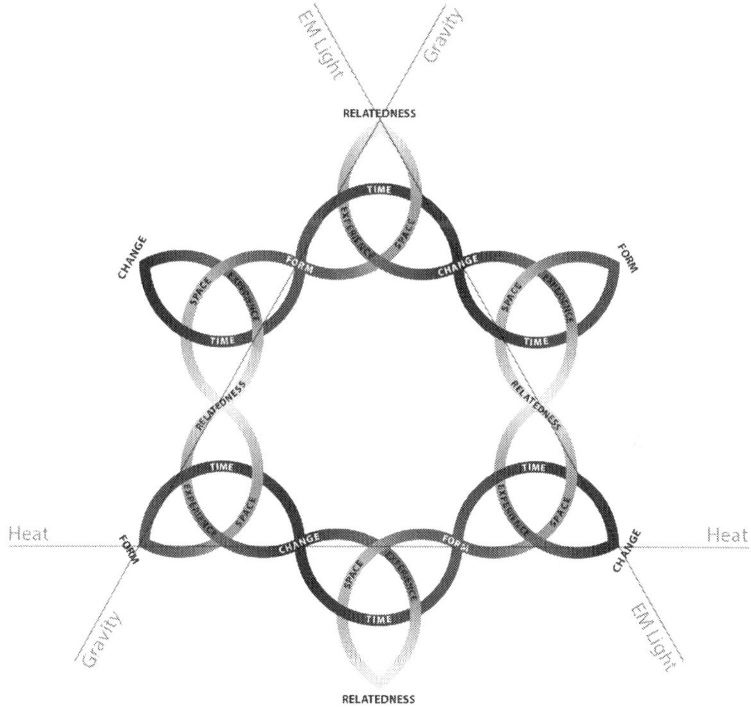

Figure 28: A stable domain with standing waves

In a stable domain (Figure 28), the six triquetrae alternate between l-spin and d-spin. The ligands offered outward from the hexagon likewise have alternating diversity of spin. This introduces some interesting properties into the triquetral spinfoam model being developed here.

The three types of triquetral wave (gravity, light, and heat) can all be identified as standing waves running through SLM states. To identify them needs careful attention, but to give it that attention now will yield benefits later. Understanding how these standing waves interact can establish anticipation patterns and sensitivities in your inner heart, the feedback dynamics of which *are* your personal values. This can iconically empower you later, and in different localities, to recognise how this model promotes substantial social wellbeing for humans. Once recognised, choices can radiate their stabilising effects out into the wellbeing of our localities in the living cosmos. Stable domains become an ecological model of the human social person, a person who is able to speak life by making life-enhancing choices in the midst of change.

To identify a gravitational wave in Figure 28, start at the bottom left red (mid-tone greyscale) form corner. Trace along the loop labelled 'space'. Follow that loop across the yellow (pale tone) relatedness ligand to the next loop labelled 'space', then on to the next red (mid-tone) ligand labelled 'form', and further on, through the next space loop to the relatedness corner at the top. This marks out a 2D sine wave. Seen in the CAD images and the printed 3D models of waves (Figures 25, 26, and 27), this sine wave has depth. It is a helix that also vibrates along its length, adding a fourth dimension of movement (4D) that becomes a measurable gravitational wave when compared relative to another. The same principles apply to identifying heat and light waves.

Figure 29 shows how a stable domain can extend when *opposite-spin* triquetrae entangle with the offered side ligands. When an l-spin ligand has a d-spin triquetra entangling, the 2D plane extends further, creating a flat sheet or honeycomb of informational processing. If a same-spin triquetra attaches anywhere in that process, however, that side ligand becomes the root of a triquetral wave extending from the SLM into a dimension other than the honeycomb plane.

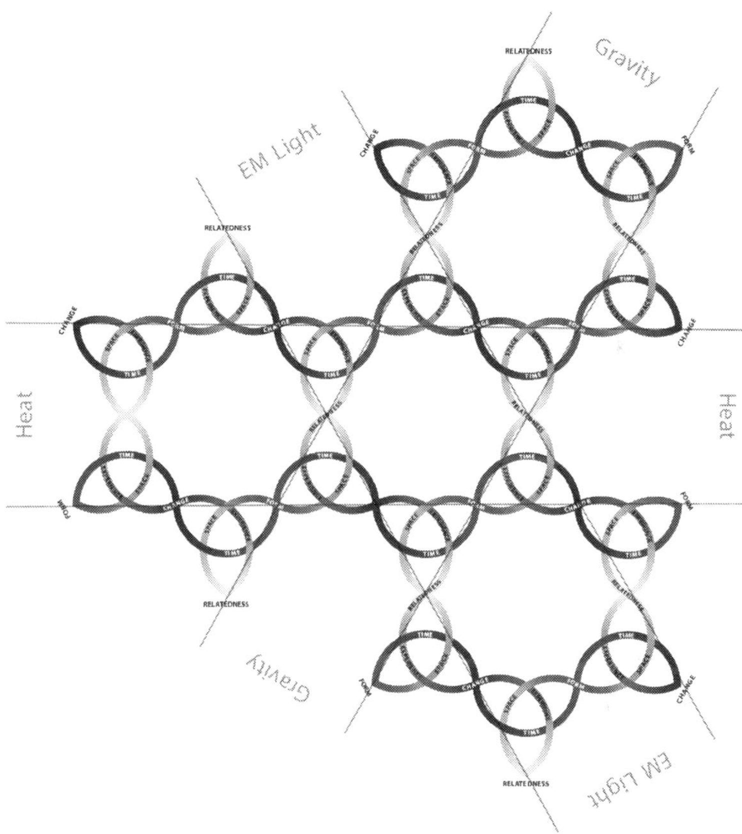

Figure 29: An extended stable domain with parallel, opposite-spin, standing waves

Notice that the standing waves run in parallel for each of gravity, light, and heat in different directions harmonically. It needs very careful observation and thought to notice a special feature associated with this. Take as an example the parallel

EM-light standing waves. If you look carefully at the upper left ligand joining them, you will see that it is a *form* ligand (red; mid-tone). The triquetra to the lower left is an l-spin, and to the upper right is a d-spin. This results in the *experience* loops of the two parallel standing waves both dipping down and rising up in the same phase. But one is an l-twist wave, and the other is a d-twist wave. The twists of parallel standing triquetral waves are opposite to each other in a spinfoam.

The l-spin and d-spin diversity is needed for stability to emerge in a spinfoam. However, as soon as a similar-spin triquetra entangles, the stable domain pattern veers off at its edge into a single-twist wave in a different 3D orientation. Because the parallel standing waves next to each other are of different twists, we may here be seeing the origin of +/− electrically charged movement, and north/south magnetically polarised movement. All of this is informational speculation at this stage of theorising, but it should be experimentally testable.

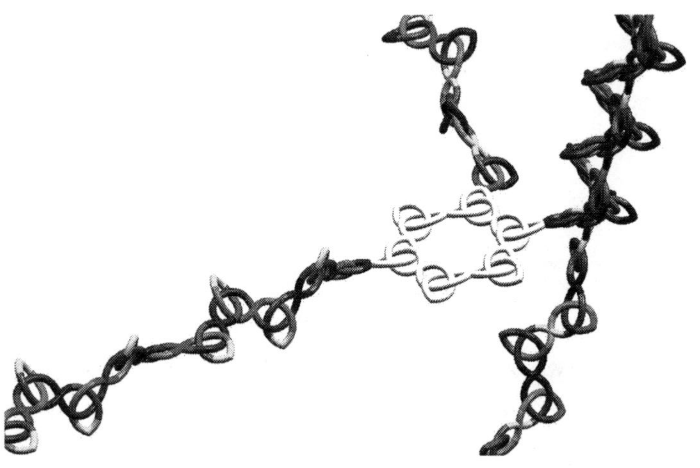

Figure 30: SLMs anchor triquetral waves in an implicate-order spinfoam

Figure 30 shows a variety of triquetral spinfoam waves spreading from an SLM, which are rooted in its stabilising influence.

To the left is an l-twist thermodynamic triquetral wave.

Top centre is an l-twist gravitational triquetral wave.

Top right is a d-twist gravitational triquetral wave at a different angle within the spinfoam.

From the side of that d-twist wave springs downwards a d-twist electromagnetic wave.

All of these, above absolute zero temperature, would be vibrating, spinning, and warping, adding several extra dimensions of informational movement into the spinfoam network. M-theory describes 11 potential spatial dimensions for informational movement. These may or may not correspond to the different degrees of freedom in the potential movements in a triquetral informenergy spinfoam on extending scales.

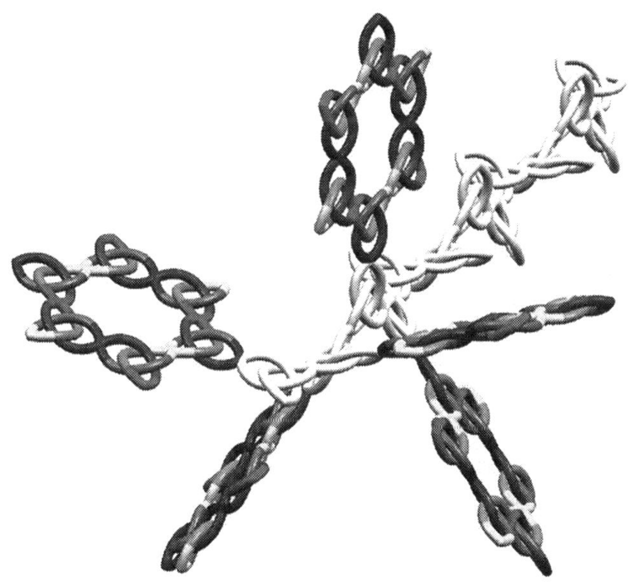

Figure 31: A triquetral wave can connect several SLMs seeding particles

Figure 31 shows how remote SLMs can become informationally connected by triquetral waves. Each of the hexagonal SLMs in Figure 31 could also be the seed of an extended honeycomb of stability in the unending movements of informenergy. If SLMs are the seeds of particles, then the human body is biophysically, biochemically, and anatomically part of this triquetral spinfoam. The potential is there for both substance and understanding to emerge from the informational dynamics that relate one locality with another, one of those localities being the memorising and pattern-recognising human brain.

Particles from topological folding of extended SLMs

In the extended honeycomb SLM shown in Figure 30, a linear helicoid wave can be traced for each type of triquetral wave. All three would interact with each other through their side ligands in a moving quantum-change spinfoam, creating a substrate for space, time, and substance, through which run standing waves of gravitation, heat, and light. Those standing waves in an SLM are triquetral synergy. The relatedness inherent to all double triquetrae converts into energy through the triquetral spinfoam fourth-dimension waves that informationally move remote localities—informenergy.

With regard to particles, imagine that in Figure 30 the extended SLM extends even further as a sheet, which can then curl topologically into an infinite array of informational patterns. Within that topology, short triquetral waves could form internal ligands of movement, which hold the overall pattern of movement within a narrow fractal range. This is a scaled-down version of the folding of long peptide chains into 3D protein molecules, such as haemoglobin in red blood cells, which can slightly shift its molecular shape when in the lungs to accommodate an oxygen atom, and then release it again in the body tissues under different environmental conditions. In this internal dynamic we can see the origin of a potentially infinite range of particles from the simple +, 0, – range of informational double-triquetral qubits. These particles would have more or less internal stability under the influence of external ligands that might extend as waves on a cosmological scale. I shall make no attempt here to map this onto the Standard Model, which would divert from the purpose of this book—to deepen human sensitivity to cooperate in shaping our shared environments.

Stable domains and their extensions may thus be the seeds of subatomic particles that aggregate their movements with more or less fractal stability into nuclei, atoms, molecules, and macromolecules, which coalesce into the membranes and microfilaments that move living cells in relation to each other, and divide and multiply into the living bodies of people who can talk to each other about how their heart-level values move them to respond differently to changing life-situations. These are the scales in which M-theory's 11 dimensions of movement probably operate. They might all be describable triquetrally.

Emergence and dispersal of movement out of nothing

Physicists have one hand tied behind their backs by the dogma that in a closed system there is no creation or dissipation of matter or energy or information. Their problem is that until 2011 the dogma was that the universe is a closed system; but it is not. Everything needs rethinking in the light of the Nobel Prize-winning observation of the measurably accelerating expansion of the universe. The cosmos in which we live and communicate is an open system.

The triquetral cosmology model can provide an account for that, which may simultaneously offer solutions to the problems of dark matter, gravity-light interactions, small inconsistencies in the Standard Model of particles, the emergence of deterministic behaviour of matter out of quantum indeterminism, and the loss of information (noun) in black holes. It can also provide grounds to discern and resist the manipulative claims of charlatans for exceptional esoteric powers, while nevertheless openly allowing possibilities for so-called psi phenomena, the unexpected diversification of life from anticipated patterned norms. This solution may be distasteful to Materialist thinkers, however, because it is a process model. There are no consistent or solid grains in it, although granularity of vibrational process at the Planck length may be a meeting point for Structuralists. In the double triquetra, Heraclitus and Democritus meet as equals. They both bring important *perspectives* on how the processing movements of life are structured, not merely imaginary. Passivity is not the final answer about how to live creatively, when responsiveness to changing environments requires a wilful balancing by choice between stative and active cooperation in life processes.

The key open-system questions are these:

(a) How does an SLM (a patterned form or a particle) appear at the end of a synergic triquetral string?
(b) And if a triquetral wave (a string) forms by attracting same-spin triquetrae to entangle from its local environment, then what happens if it never does encounter the opposite-spin triquetrae that would start a stable domain formation?
(c) And finally, where do isolated triquetrae come from anyway?

There is one simple but philosophically precise answer to all three open-system questions. The *probabilities* of these potential events depend on their local and extended holomovement *environments*. And here is the unexpected... an environment may be *nothing* if it is unchanging.

According to triune principles of organisation, a single triquetral quantum change has no existence on its own. Change only gains existence when it entangles with another to produce a resonant, informational, vibrational synergy. Otherwise... nothing. Given these considerations, however, the principles of organisation of movement are alive prior to the emergence of existence.

***

Let us explore the implications of this widest possible context—the process of progressive emergence of existence out of nothing—by revisiting those three open-system questions.

The answer to question (a) has already been given. Let us imagine that a triquetral l-twist light wave extends into a changing triquetral environment, even one where superheated double triquetrae fly around like flickers of flame at a subatomic level. Some of these may even break apart into single triquetrae, which then lose synergy, shedding mutual informational change. They might fade out as they lose their spin, and cease to exist. Other single triquetral changes entangle with the advancing light wave, adding to its advancement as a triquetral spinfoam wave if they are l-spin; until a d-spin triquetral change entangles with it instead, which levels out the twist into the seed of a stable domain. This forming SLM may extend if it gathers more alternating l-d-spin triquetrae. The resulting seed of a particle could, so to speak, anchor that end of an advancing light wave.

Given the harmonic or resonant nature of the entire spinfoam, the emergent stability of that SLM or particle might reflectively enable the entire triquetral pattern of waves to become stable enough to absorb and transmit more informenergy changes as higher-dimensional movements along its length and reflecting back as a fourth-dimension change event (a particle). Every action has an equal and opposite reaction in the explicate order, emerging possibly from the bidirectionality of holomovement.

That fourth-dimension energy wave would be slowed by any side ligands forming into a spinfoam along the advancing triquetral light wave (or any EM wave). The spinfoam might thus have the stability to *radiate* the light, *because of* the creation of a new particle at its end beyond our everyday space-time mental framework. If, however, there are no d-spin triquetrae to entangle, then that advancing end of the l-twist light wave might become unstable. It may in all probability be shaken apart by a fourth-dimension energy surge, which might break the relational synergies apart, extending the collapse back to the nearest stable domain, which may itself transform into instability in a change event known to physicists as particle decay. Such a decay may be measurable as, for example, nuclear fission. That would be one possible answer to question (b).

Another possible scenario would answer question (c). If apart from a spinfoam of changes there is nothing, then the spin of an advancing l-twist light wave might itself be able to induce a potential change in its empty environment of nothing. To picture how this might happen requires two Zen paradoxes, which might take several years or even decades to appreciate, but it is worth the effort to start now because of the potential benefits. The first paradox is to try to imagine a dot that is both the centre and the expanding circumference of a circle. That is difficult, but not impossible to imagine, if you include movement. It represents the growth of potential within nothing, which may be induced by a changing environmental quality of relatedness in that movement.[1]

The second paradox is more difficult to imagine. It is a question. What happens when the unstoppable force meets the unmovable object?

It is worth pausing to consider that, perhaps for a few years if necessary. Both features must remain true.

I am going to short-circuit this personal development reflective method, however, by offering a possible solution. This offer may be visually imagined iconically by considering the two potential spin types of a single triquetra. Look again at

Figures 5 and 6. Imagine a circle's circumference (which is also the expanding dot at its centre) developing three equidistant further points—three being the necessary and minimally sufficient number to generate diversity in unity. Imagine a continuing expansion of the three arcs between each pair of these points. They now start folding 'inwards' and twisting, creating an inner phase space. The expansion continues so that those arcs have become semicircles with their own inner curves as sine waves. Each is therefore a curling half-spin of change, connecting two of the three diversifying points on the dot-circle's circumference. This curling expansion thus turns a 2D dot-circle into a 3D pattern of movement. It then becomes a 4D phase space of cyclical changes, its criss-crossing lines including time as they make a fractal pattern of inner movement.

If you were ever to try doing this with wires, which you have cut and curved into semicircles connected by loose hooks and rings at their ends, you would soon discover, as I did, that one loop of wire *has to pass through one other* in order to make that 4D phase space. The paradox of the unstoppable force and the unmovable object now becomes relevant. Informational integrity within the expanding and curling dot-circle *has to remain* while a triquetral spinfoam extends in a self-stabilising way out of nothing, to emerge into an informationally united cosmos that sustains shared life.

In this *process cosmology,* nothing is 'unmoving' (which is a paradoxical statement in itself, its opposite pole being 'everything is moving'). In the Zen paradox, the unmovable object, however, *exists.* Existence follows from a triquetral cosmology process. The stability emerges through co-origination from three diversifying points on an expanding dot-circle. An unmovable object exists therefore only as a *stative process.* Existence is not static. 'It' (the object) does not have a static core. In this Zen paradox, the unmovable object is a highly stable triquetrally patterned form. When an unstoppable triquetral force encounters it, however, the triquetral movement of the force is itself a fourth-dimension movement through the same spinfoam that is only patterned differently. The wave pattern can therefore itself transform into the stative object's pattern. The stative object's pattern is a tensioned whole. It can therefore hold and transmit triquetrally patterned waves on its far side as an unstoppable force. The unstoppable force has thus moved informationally on through the unmovable object. One has transformed *relationally* into the other.

The creation of matter and existence out of nothing might philosophically be a possibility if this process of triquetral expansion is set in motion when a spinfoam extending wave becomes an environmental *potential in nothing*. Quantum change may then emerge as a relational search for transformation. If several triquetral waves are advancing, they might make a complex environment of potential. This would add a degree of uncertainty or unpredictability to the spin type of any quantum change that might emerge in this expansion. As a consequence, whether waves or particles appear out of nothing is determined by probability in relational qualities of the environment.

If this process could happen everywhere (or anywhere, variably depending upon diversifying environments), then the conditions exist for an open universe to be expanding from within at an ever-increasing rate.

***

To round off speculations about the physical universe to which triquetral cosmology may add new potential, a few ideas and loose ends follow. Dark energy-matter makes up 80–95% of the energy-mass of the universe. Some of this could be a configuration of SLMs that are connected by opposite-twist waves to those that our scientists' instruments measure. Small inconsistencies in the Standard Model of known particles may arise from partial SLMs that are highly unstable. Deterministic behaviour of matter may emerge out of quantum indeterminism when extended SLMs have high levels of internal stability. And the loss of information (noun) in black holes would not be so mysterious if informational patterning is sourced in a triquetral spinfoam process, through which the measurable wave-particles transmit as a fourth-dimension energy. That spinfoam can both form out of nothing, creating expansion, and return to nothing as isolated superheated triquetrae in black holes. Informational order may disperse as this process moves on in its diversifying way. Black holes may, perhaps alternatively, degrade order into the isolated or double-triquetral changes that may temporarily prime the darkness with a substrate for further entanglements. This process would constitute the reordering or renewal of life.

The power of choice in the cosmos

However much or little this triquetral hypothesis says about the physical cosmos, this book starts from the neuropsychology of the power of human choice, aiming to refresh the philosophy of movement that may follow from making decisions between options for behaviour and communications. Conscious mind holds the middle scale. Living society emerges that has enabled an agreed approach to scientific exploration of the micro and macro cosmos we all share. This book's focus is on how to live creatively or regeneratively when physical and social environments crumble, as they are about to do.

The key concept being developed is that *effective relatedness shapes the future substance of life*. When considering movement, some people naturally ask, "Movement of what?" The triquetral approach replies, "What of movement?" Future substance can emerge and shape from living the relational process well. 'Well' here means intra-actively with the right attitude, which is the subject of Chapter 12.

- A Materialist or Structuralist thinker will tend to imagine the spinfoam quantal network behind substance as some thing that is there to be observed. In a conversational frame of mind, however, there is movement. Conversational movement is more like a word spoken person to person, which can be heard and responded to, having been received into a balancing process filtered first at heart level.
- An Informationist or Individualist thinker will tend to imagine how it is possible to use the spinfoam as a medium through which to spread their information (noun) to get what they want. In a conversational frame of mind, however, there is environment. Conversations move with bidirectional informational synergy that calls for a heart-level response, out of which new understanding emerges, and possibly new substance.
- A Vitalist or Collectivist thinker will tend to imagine that there are trends of movement in the spinfoam that can be discerned by divination, to the authority of which they will need to submit in fearful appeasement for a healthy eternal future. In a conversational frame of mind, however, the eternal now is a bidirectional relationship of life that requires the choice be made to engage constructively or creatively in the changing circumstances of life with others.

In the final chapter we look at intra-action. We look at how the social and inner ecologies emerge in a physical world. We look at the group dynamics that stabilise or disrupt physical environments. We consider how the potential of the land to renew life's order and future creativity is affected by the qualitative ways that people choose to interact with each other.

Chapter 12

Movement Shapes Our Social Ecology

Triquetral informenergy extends into e-motion

A picture has been presented that substance—the physical environments of land, seas, biosphere, and space—moves relationally and shapes into its forms because quantal changes of informenergy synergise relationally within. This is a picture of how ordered life as integrating movement dwells within substance and can emerge from parallel processes in the physical brain as understanding.

In a triquetral view of life, there is not a primary chaos. There is primarily ordered movement that can be reduced to chaos under stressful circumstances, and can recover from it. The dualistic view of external forces acting on objects can now dissolve into a triquetral vision of depth movement. Forces and objects are both informational processes. They are differently patterning in their complex systemic relations to each other. They are mutually synergistic; they are not distanced and separated.

Synergy, in this quantal-change picture of life, is a precondition for energy and for mass. The speed of light in this moving picture of life, which connects the repatterning of energy and mass, is a product of networking in the underlying synergic spinfoam. That same networking of movement in quantal spinfoam creates the conditions for living forms to emerge, and also for these living forms to become aware of their mutual connection into synergic repatterning through their behaviour.

Fourth-dimension waves of movement through this quantal spinfoam become measurable substances and forces. Comparisons can be made wilfully between human experiences of them, creating the impression that substances and forces are self-existent, when in a triquetral reality they are movements. Measurements become more 'accurate' as agreements change to choose a better standard for comparisons in subsequent conversations, for example between traders and customers, or between scientists in competing teams to compare their experimental results. The same quantum-change spinfoam that gives substance to energy-matter gives that understanding to the social mind.

We now move on from physical substance to how informenergy emerges in the inner human experience as an understanding of social ecologies. The capacity to make life-enhancing choices gains synergy by first letting go of dualistic values. Mutual understanding may follow a deeper awareness of the order within movement, which can break apart and get stuck as people fix their minds on something objectified. Letting go of dualistic values does not mean abandoning the value of each analytical perspective on life, however. It means that greater value is obtained when our histories as Observers of life, Creators of life, or passive Participants who receive life are brought into the full conversational adaptability *that unites diversified life*. With these analytical perspectives, we become intra-active partners in adaptive life together. The three ecologies become one diverse synergy when differences can be celebrated for the life choices that they enhance.

Mind and matter meet in the human inner heart. Awareness of changing patterns of relatedness can emerge first below the orientation horizon. Neurological networks are informationally transformed by biochemical and biophysical sensitivity to environmental movement. Awareness is not merely a mental representation of those movements. Awareness is a fractally scaled transformation of movements into another informational medium, namely the neural network, which is electrochemical and structured by synaptic connections. Synapses in a neural network are the informational equivalent to triquetral ligands in a spinfoam network. The principles of informational modulation of synergic movement by remote 'sideways' influence on their function are identical.

Fractal scaling of patterns from the environment into human neural networks feeds the core limbic brain first. The core limbic brain intimately associates its neural patterning with hormone releases into the body's bloodstream, and in this way coordinates the whole person's (mesodermal) living tissues. The whole embodied person is physically repatterning *with* the core limbic brain.

In the core limbic brain, sensory feedback from the social and physical ecologies meets or clashes with anticipations and values from the inner ecology. Mismatches or satisfactions scale up into a generalised emotional state of affective awareness, with an associated corona of active emotional preparations affecting the body's muscular poise and facial expressions. Genetically, emotions are thus visible for all to read and potentially understand. They alert the social ecology to change. The life of the social ecology therefore, or its deathliness, depends upon other people's sensitivity to revealed messages about personal values, which people may choose to keep hidden in their inner hearts.

In this personalised and diversified way, triquetral informenergy extends from the physically ordered substance of the land into the inner order of the body and mind, and simultaneously into the social ecology, unless blocked internally.

This inner transformation and adaptability of informenergy is shown iconically in a double triquetra as the curling inner loop at each end of the connecting ligand. Dualist yes-and-no-no thinking, and no-and-yes-yes political comfort zones disrupt triquetral informenergy by breaking that inner loop, veiling the creative feedback cycle of engagement in life. They break the synergy. Synergy and creative adaptability can be restored conversationally, however, potentially reshaping the energy and mass of life. The choice is ours.

The presence of loving kindness is the core human state

Choosing to open an occasional gap in time for reflection can help to restore conversational orientation by recognising and integrating previous dualisms that divide personal and group identities. The aim is to explore a yes-yes-yes triquetral balance of the embodied human nature. This is revealed, as explained above, in the balanced poise and action plans with which we move life. It is not sufficient to retain that balance as a purely mental state of the inner ecology.

The embodiment of human nature includes, and does not exclude, the biophysics and biochemistry of the scaled-up triquetral standing waves identified in Figure 29—light, gravity, and heat. These are active within the macromolecular transformations of every living cell of the body, including the pattern-recognising neural and immune system networks. These pattern-recognising networks are informational systems that connect the substance of our personal identity (our values that incline our responses) with the wider social and physical ecologies. This physical connection enables us to survive and thrive and adapt in different environments. However, we are not aware of physical light, gravity, and heat in our interpersonal dynamics as physical forces, even though they are there. We tend to take a philosopher's dualistic step back from them, and call these physical experiences something else. We need to step forward into a healthy gap in time if we are to see triquetral parallels between the 'language games' of science, philosophy, and humanity. Let's take the risk.

- *Electromagnetism* describes scientifically the changing relatedness that we are aware of as *substance*. The equivalent informational process in mindful engagement in life is *understanding*. We name the relational frames of concepts, which are bundled feedback loops of electrochemical activity that have become synchronised within neural networks. We decide whether they attract or push apart from other concepts, and thus categorise our experiences in left and right hemispheres. Our inner understanding may then possibly synergise with another person's in a conversation as we enter a social environment. This mutuality may lead to us experiencing a sense of the *presence* of another real person. Therefore, a change-relatedness-change-relatedness triquetral wave could represent a sense of *presence* influencing emergent understanding in a social environment, as much as it also represents a light wave (an EM wave) in a physical environment influencing the potential emergence of formed substance.

- *Gravitational* triquetral waves similarly describe scientifically the relatedness of forms that we know as movements in space. The curious feature of these is the *enduring connection* they establish between forms everywhere, even while light and heat strain the gravitational attraction and even push objects apart. Enduring connection in a social ecology is commonly called *love*. Associated with hope, love describes enduring connection despite change, perhaps even leading to reconciliation after brokenness of relationships. Therefore, a triquetral wave of form-relatedness-form-relatedness could be used to represent love iconically in a movement that structures stable domains in the social ecology, just as much as it also represents gravity.
- *Thermodynamic* triquetral waves likewise describe scientifically the changing forms and evolving states of energy-matter. Our experience of changing forms allows effective timing of choices for relational interventions,

action, or rest. *Responsiveness* of movements to environmental change is core to thermodynamics in materials and to the timing of personal choices. In materials, that responsiveness of forms brings about the progressive spread of heat, and its dissipation into radiant light. In the social ecology, the equivalent responsiveness is the *warmth* of social interaction, which may spread through a group shown in timely and merciful acts of *kindness*. Therefore, a triquetral wave of change-form-change-form could be used to represent responsive *kindness* iconically in a stable domain of group dynamics in the social ecology, as much as in materials in the physical ecology.

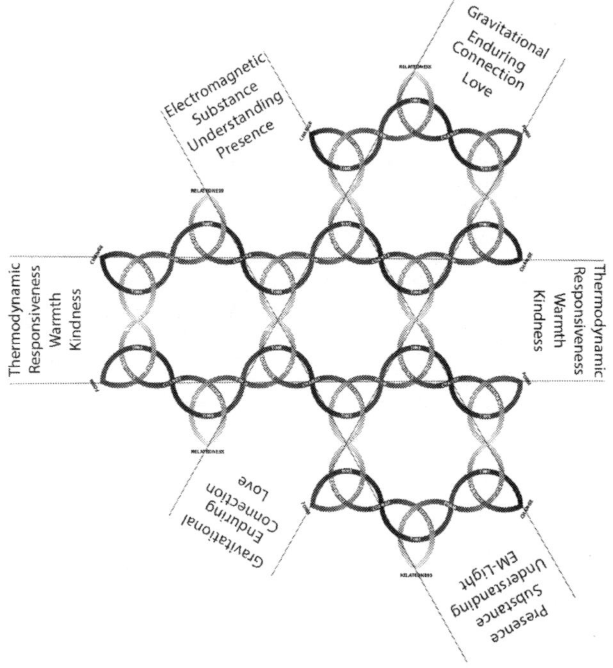

Figure 32: A substantial social ecology is the presence of loving kindness

Figure 32 brings together the triquetral parallels in these different 'language games'. It extends Figure 29 from only the physical ecology, showing how a social network thrives when integrating the same triquetral principles of organisation that pattern movement everywhere. In Figure 32, each embodied person is represented by a single triquetra, relationally responsive with others, and each having their unique emotive experiences of change that depend on memorised histories, their anticipations, and their sensitivity to their local environment within the wider network. Psychosocial diversity enriches the underlying unity of social connection.

In a healthy social ecology, each person uniquely can be the presence of loving kindness for others. They might behave and make choices as if the other person were as significant in the overall picture of life as oneself. In fact, others fill out the individual's personal identity with the character of their chosen qualities of relatedness. Putting oneself down, or apart, or becoming persistently self-questioning or self-critical or subservient is not a triquetrally balanced state to aim for, although all of these may be phases in a process of seeking mature balance. On the other hand, mutual respect, listening to discern personal values, hearing another's inner heart, naming personal values together, adaptability to welcome others who differ, hospitality, celebrating together in nature's rhythms and cycles, remembering our roots in a unity of movement... all of these are features of creatively belonging in life's synergies.

Hurts, setbacks, disappointments, and change can turn the presence of loving kindness to grief, which is not the end of love or kindness, but a more dynamically primed mode of love's enduring responsiveness, moving people restlessly to explore how to reconnect when change has pushed them out of a comfort zone. There is a realistic risk, however, that the unpleasant emotions of love's grief mode will additionally be misunderstood, and push or drag people to withdraw their presence, and to additionally lose warmth, perhaps temporarily. This could be an honest phase of healthy adjustment and adaptation in deep bereavement, which calls for compassionate action from others to sustain connection. It is all part of maintaining balance during times of change within a social ecology that has the potential to thrive and encourage recovery, restoring the capacity to be present in a social ecology with renewed loving kindness.

Friendship shapes the local substance of life

Kindness differs from love, in that love's main modes of joy and grief represent stative and active states of enduring connection during times of change. Kindness also has its modes, as soft and hard states of relating. Kindness may need to encourage or motivate responsiveness in others who have become unresponsive socially. So too can presence be in modes of local and remote. Remote presence is the continuing sensitivity to informational 'word' embedded in spinfoam synergies. Remote words can sustain heart-level awareness and valuing even when for a time the pressures of life have separated people.

The core human state thus has plenty of scope to diversify into a complex state of sometimes confusing unpredictability and uncertainty. There is a relational state that can seed order in such potential for chaos, where patience allows renewed understanding to grow. That state is called friendship. This relational quality, extended also to people who differ, can open the door to the restoration of life and thriving as the innate order of life's movements re-emerges from the implicate order.

The relational synergy of friendship can enter any type of human relationship, and depart from it when trust is broken. But few would doubt that where friendship is part of life, the substance of life shapes up differently from where it is absent. Synergy affects thoughts and substance in equal measure; lack of trust likewise.

Where the land is cared for, its fruitfulness increases. Choosing to seek or explore synergic friendship with those who differ is potentially life-enhancing. We can cultivate each other's hearts in rather the same way that a farmer or gardener can cultivate the land. Where friendship extends to love of the shared land also, all three ecologies unite at an informenergy level in a harmonic or resonant interrelatedness. Synergy between people, who all emerge from the land, may extend its effects in a reverse direction also, coordinating a similar effect even on the land's fruitfulness. All equations in the maths of physics are equally true in reverse. This effect on the land has been recorded in a number of settings around the world in a video series called 'Transformations'.[1] The point of these journalistic reports of the impact on the land of qualities of human relationships is this. At a profoundly deep heart level, when dualistic thinking can be set aside and qualities of relatedness are understood to be formative, humanity is earth organised by the same principles that govern matter, and matter has within it the potential to manifest life.

Exclusive group identities can corrupt humanity

A profound problem that humanity faces, however, is the way that a shared mental focus of attention can quickly corrupt synergic friendship. Exclusive group mentalities can emerge in the spinfoam of life. The self-interested focus of cerebral mental attention can drive conversations, and when people see life in a similar way the thoughts of their heart may incline into a locally synergistic analytical frame. The heart-level value system is still involved in this inclination. However, without insight into the problems that may follow, exclusivity can enter the conversational equations, aiming to remove distractions created by those who see life differently and who are difficult to understand. Diversity, subtlety, and nuancing of ideas can be squeezed out to protect a status quo. The person may feel empowered to enhance their own self's identity in these restricted, closed-system types of conversation.

A solution to this core problem may emerge at a heart level, prior to reasoning, if this narrowing process can be pictured iconically in a memorable way. Then people can recall the imagery and protect the fullness of their self and each other in their various, healthily diverse social groups, by encouraging friendship to remain open to those who differ.

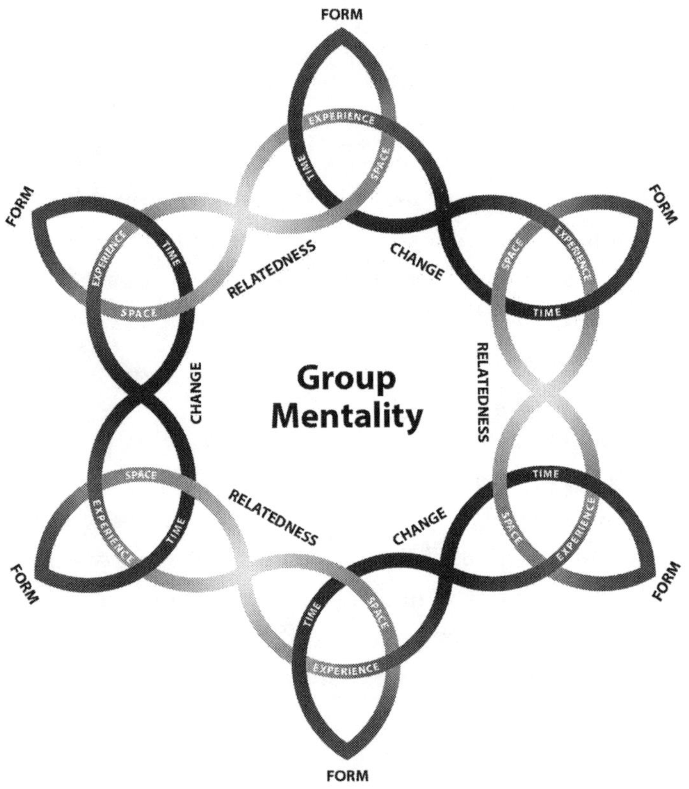

Figure 33: Restricted group membership—group identity

An icon of restricted group mentality is shown in Figure 33. All the external ligands offered from the social group are of the same type, which differs from Figure 9 (Group identity allowing diversity), in which all three qualities of connection are offered by the ligands extending outwards. In Figure 33 the restricted group would accept only those who have a particular form; for example, members of a classic car club (red; mid-tone in greyscale) need a classic car to join the group. Figure 9, by way of contrast, invites diverse people to join the group in friendship. In an iconic stable domain such as a social group, the diversity arises from the ligands in the 2D plane having a spin that orientates diverse ligands outwards from the hexagonal ring. If some of those internal hexagonal ligands were to spin the next triquetra 180 degrees instead, before it fixes into the hexagonal stability, then the inner diversity would be reduced by turning what could have been an outwardly directed diverse ligand inwards instead. In this way, only two of the three principles of organised movement recur alternately in a forming ring. The ring effectively becomes a closed circular wave in 2D, a reduction from the creative spinfoam array of waves and stable domains. Scaled up into social group mentality, three types of restricted group mentality emerge that display the characteristic effects of each of the triquetral waves in matter.

Exclusive group mentality imposing form

Figure 33 shows a restricted group identity that imposes conditions on form to join. Other examples are cultural, racial, or faith groups that require certain forms of behaviour or belief to conform, commonly being shown in the use of uniform clothing, badges, or insignia such as salutes that help people to identify themselves with other members of 'the club', or the gang, or the special-interest gathering, such as car clubs or sports teams. Those in the group feel bonded by a shared sense of their presence informationally through that structuring formation. It boosts their own identity and feeling of worth or value by excluding those who differ. People feel enlightened by being in the group, and perhaps can feel empowered or self-justified when they see those excluded as being shut out in darkness.

To describe this feature of humanity is not to criticise it. All humanity benefits from having a sense of belonging with shared interests. The purpose of mentioning this is to increase sensitivity to the risk that this can corrupt into 'belonging by exclusion', which would impact the wider creativity of life by spreading a wave of grief rather than joy. Once aware of its wider presence as a subtle shift of informenergy level, group identity can be greatly enjoyed *without decreasing the value of those who are not members of it*. Historically, this awareness led to an emphasis on hospitality and welcoming strangers as if they are part of the group or family for the timeframe of their presence. Without this cultural emphasis, the perceptions of others who are not in the group may build up resentment and opposition, with competing claims for self-justification of diversity. Enhancing the grief of others, without clearing the way to live with a shared sense of belonging in a higher order of life together, will destabilise the future.

Exclusive group mentality imposing change

Figure 33 could just as easily have shown a different type of group characteristic that has a purpose of imposing change (blue; dark greyscale) on chosen places in physical or social environments. The group will share some love of a particular feature of life, and strive to impose that attitude on others not in the group. Examples are political action groups, animal rights groups, evangelical religious groups, environmental concern groups, healthy living activists, and so on. They strive to change the behaviour of others, mostly for very life-enhancing reasons, but not always.

The problem comes, again, when the sense of belonging and group empowerment subtly shifts to an attitude where those outside the group are devalued or heavily criticised as unloving. By reducing diversity within the group to increase the effectiveness of sharing the informational message, the group may unwittingly aim to reduce diversity outside the group also. This mistake is driven by a false and reduced mental concept of perfection.

This problem for group members could be overcome if they were to acknowledge the subtlety of depth informenergy communications that enhance daily life. People's hearts are moved widely, below the orientation horizon, by informational sharing that is perceived intuitively as coming from life-enhancing values. Awareness of this intuition can be resisted, however, when an analytically dualistic perspective filters out certain sensitivities. On the healthy side of conversational reality, loving life in a way that includes diversity has a subtle effect of moving or transforming other people's hearts by raising awareness. Curiosity above the orientation horizon then introduces the potential for change. If members of a pressure group hope to have a life-enhancing effect, they need first to enduringly connect at a values level, so that others first feel heard and may then be willing to listen to a clear expression of values. Informenergy transforms life from within at the personal reality level of values. External pressure on behaviour then becomes superfluous when the core-level informational sharing is based in loving kindness.

Exclusive group mentality imposing qualities of relatedness

Figure 33 could equally show iconically how the third type of restricted group mentality offers only relatedness ligands outwards (yellow; pale in greyscale). This one has a wider purpose of cultivating a warmth of relationships over timeframes, or in a timeframe of eternity. Examples are religious or philosophical groups, commercial sales teams using sales social-psychology, and humanitarian aid or support groups.

While based in kindness, the methods by which these relationships are built may lead to hidden problems. They may induce a 'needs led' quality of caregiver-receiver or provider-recipient, rather than a values-based equality of mutual respect. Subservience and dominance may enter the equation, so that people who have initially found their life quality improves through relatedness then find they feel trapped and obligated to conform in some other way.

The solution is to ensure that the model of perfection among care and humanitarian aid providers, sales forces, and spiritual leaders or guides is to liberate people into their own relational exploration of life. Freedom does not imply isolation and disrespect for relatedness, or shaking free from others. Freedom to relate constructively and creatively with mutual respect is life-enhancing in changing environments.

Attempts to balance a moral philosophy

Given the complexity of the social ecology, and the associated inner complexity of personal development and growth to maturity, it is not surprising that philosophers over millennia have formulated moral codes for living to guide personal choices. Difficult situations will always arise, usually unexpectedly, requiring a balancing act of values when making a choice between life-enhancing options for active or stative responses. Triquetral cosmology may offer a way to balance even these diverse moral codes.

Duty-based and utilitarian love in analytical perspectives

Historically, there is a broad category of two types of moral code. One places an external duty on individuals to behave in ways that support a wider environment. These are called deontological ethical or moral codes, from the Greek *deon* for 'duty'. The opposite category is called utilitarian, from the Latin *utilitas* for 'use'. These moral codes focus on the freedom of individuals to find a life-enhancing path through the complexities of life. They put a duty on others to allow those freedoms, claiming the utilitarian purpose that 'the end justifies the means to get there', even if others disagree from their viewpoint.

Needless to say, there are enthusiasts for each category who raise justifying arguments to support each perspective. If adherents to each would frame their choices in love, seen as enduring relatedness, and with qualities of kindness made substantially real, and bringing their presence as a quality of the shared environment of society, then a balance could be found between these opposite moral codes, deontological and utilitarian, in any situation. Usually, however, people are not seeking a balanced life. They are seeking a comfortable or safe one, free from suffering. If so, the arguments about 'intrusion' become more divisive.

A triquetral balance of ethical principles

The issues come into focus when considering practical life-situations that require a decision. *Situational ethics* has been traditionally claimed as the focus of utilitarianism, opening the way to needs-led thinking. This has been contrasted with the deontological view that all situations need to be seen in the light of a bigger picture that sets priority values on individual needs. However, disagreements then follow about the nature of the big picture, and about the nature of needs or wants, which can cripple a moral response when facing a difficult situation where life or death or disability of the self or of significant others (or of 'insignificant others') is called into question. Denial of responsibility, of course, is an easy path out of dilemmas. However, that is not the path that enhances life when shared environments are crumbling and forcing change on everyone. A more robust picture of balance is situationally needed, which can adjust appropriately to prioritise factors that influence the complexity of changing situations, especially where large numbers of people are simultaneously affected.

Fortunately, the icon of movement used in triquetral cosmology can also provide a balancing bowl in which to hold and gently swirl around options for behavioural responses, while bringing heart and mind into a creative partnership for choice-making.

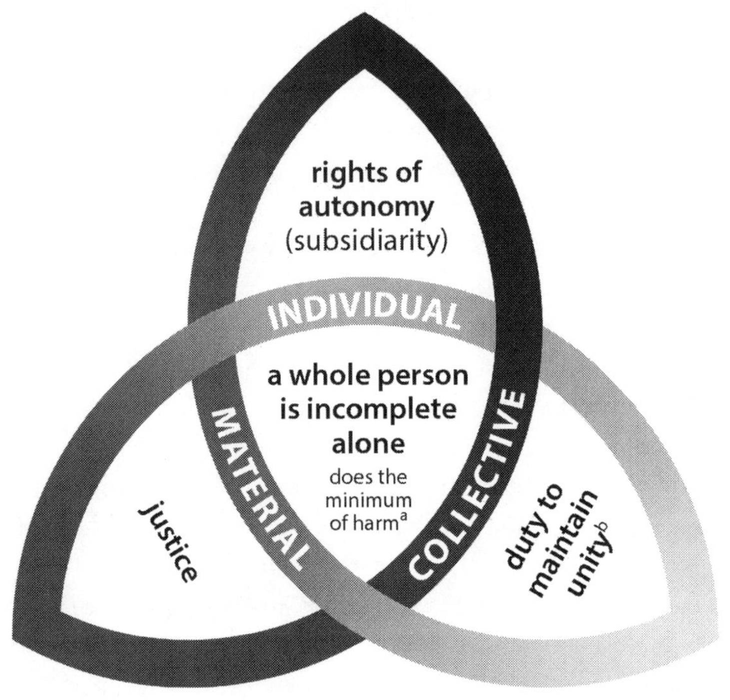

Figure 34: Balancing ethical principles in difficult situations
[a] Non-maleficence. [b] Beneficence. Beauchamp & Childress, 2013

Figure 34 maps the model of medical ethics presented by Tom Beauchamp and James Childress onto the triquetral principles of organisation.[2] In theory, these principles apply to every level of the material and social cosmos. In the central phase space, where timing is integrated, the whole person who is incomplete alone (the double triquetra is the core icon) is perceived as an integrated presence of loving kindness, unless problems have blocked that balance. Wholeness is then fulfilled when curiosity adds qualities of relatedness to explore the values others share equally with oneself. Overcoming assumptions with curiosity is the ethical quality relevant to making decisions in this state of indeterminism. The first priority then to maintain balanced principles of organisation in the social ecology is to do the minimum of harm (technically called non-maleficence). Obviously, that is open to interpretation, but by bringing a lively understanding of the intuitive heart and the way rational thought is inclined by its values, a sensible and sensitive conversation with oneself and others can discern the extent of potential harm within different choice-options in situations. Making potential loss lists, as in the Emotional Logic method, is a useful practical tool by which one shared personal value can be selected at a time to agree action plans that focus on building a future that sustains that feature of humanity.

In Figure 34, around that central phase space are the three analytical viewpoints to balance in moral choices, called the Individual, Collective, and Material perspectives in this setting.

Individual adults do have rights of *autonomy* within civil society. Responsibility must be devolved from central authorities to *subsidiary* organisations in which people have roles to fulfil, so autonomy is always regulated in society. Nevertheless, people need to learn how to exercise responsibility (responsiveness) in their various roles, so that they bring life-enhancing changes into their environments. Children need to explore and play at making autonomous decisions and reviewing their impact in safe settings, otherwise they will never develop into mature and reasonable adults, or capable parents. Without feedback cycles of review by their parents, autonomy in children can lead later to dictatorship, selfishness, and hard-heartedness in adults. So, autonomy is vital to bring unique personal capacities into play within society, in an adaptable balance with the informenergy brought by other complex factors.

The Collective perspective brings a sense of duty to maintain unity so that 'the people' as a whole can thrive, the land also potentially thriving with our ecological qualities of relatedness. This is sometimes called 'doing good' (technically called beneficence). For many people in individualistic societies, this perspective has gained a pejorative sense of 'do-gooders' interfering in the lives of autonomous others. Nobody said that finding a balance is easy, but to have an iconic image that draws attention to a third value (justice) can ease the strain on these competing two. Beneficence has the obvious problem that individuals are in no position to know what 'good' means for another person to thrive, or even for the whole of society to thrive. But having the concept of 'life enhancement' as a measure of 'the good' can facilitate a balancing conversation. And further, balancing the individualistic and utilitarian focus on reducing *unique feelings of suffering*, by introducing the language of personal values and losses, opens a wide new range of conversations that get behind the veils of dualism. Personal values are the lived reality of the social ecology. Many mature people are able to live creatively with suffering and without complaining, because they know that living by values that are named and shared is more life-giving to the whole; and by social feedback more self-fulfilling.

The Material perspective brings into focus the substantial and structural balancing of life. The distribution of resources in times of want or need can be a serious source of dispute and conflict between individuals and between groups. Power dynamics come to the fore, with oppression and deathliness stalking as the enduring connection of love turns from its joy to its grief mode. People may try to render others powerless to change life by enhancing their grief.

How individuals and groups will respond to the serious disruption of environments and loss of structural resources in the next 30 years is beyond predictability. What counts as justice will depend on perceived personal and group values. And this returns our attentiveness, from the dynamics of dualistic and broken life above the orientation horizon of humanity, to the heart-level dynamics of sensitivity, awareness, and intuitively inclining the thoughts that emerge from movements of the inner heart.

Justice can only emerge from the mercifully responsive heart of a person who is willing to be substantially present with others with an attitude of transformative loving kindness in their understanding of people's needs, wants, and searches for empowerment. A triquetral balance in the heart can incline such movement. Life-enhancing informational exchange through the mutual balancing of energies may hopefully result and transform the explicate order. Meanwhile, in the implicate order, the underlying synergy of continuous creation may be allowed then to bubble to the surface again in a renewal of life.

Recognising the primacy of synchronous synergy as the reality of relatedness is the starting place for transformative living. Relational qualities between people become more resonant when appreciated through a gap in time. Fruitfulness of the shared environments of land, sea, air, and space can also synergise and renew when given the opportunity by human choice to do so. Naming personal values, therefore, is the business when connecting into peace in a gap in time to thus view states of relatedness. There, endings can turn into beginnings of a new story of life.

Epilogue

The Renewal of Life by the Call of One Who Loves Us

So begins a journey to equip thoughtful people to know how and why a compassionate response to others in the current environmental climate crisis would physically change the future more than making a grudging one.

Contributing to the problems that beset a social ecology is the way that individuals may prefer to focus their attention on soothing or avoiding their own unique emotional feelings without *first* meaningfully attending to their informational content, in terms of naming their underlying personal values. They can shift towards a healthier state by choosing to create a small gap in time for reflection. Unpleasant and even overwhelming emotions can be untangled there, if attended to by fitting them into the Emotional Logic mental framework.

This framework transforms challenged emotions into the growth of new responsiveness. Unpleasant emotions can be mindfully engaged without pushing them away into Denial. Their informenergy can be recognised, absorbed, and understood using the mental framework that directs thoughts back to the heart-level values that these emotions evidence. By thus bringing mind and heart into a creative partnership, an ecological wholeness of the developing person results. At heart level, this wholeness engages into an open system of life shared with others and with the land. Rather than seeking separation into a dualistic mental or spiritual realm, this method naturally disperses the suffering caused by unpleasant emotion by making choices to explore how these emotions can fulfil love. They are not an end of love, but love moved into a more dynamic mode to energise mindful adaptability in a wider open system of values and responsive relationships. That awareness and choice for movement in the social ecology can start to heal not only the inner heart of humanity, but also through humanity the land too can be healed.

And finally, your inner heart is potentially an open system for you to hear the call of one who loves you. Awareness of implicate movement is an awareness of Infinite Love. Comparative consciousness can grow from that root within individuals as the inner heart grows in personal strength, even while the closed systems that people have created are collapsing all around. You have the potential to contribute to the synergy of that call to renew and restore Life. The choice is yours to move into that place of peace and fullness.

Notes

Introduction

1 In 1991 Doc Childre founded the HeartMath Institute to study and teach how the biophysics of the *physical heart* is more than its biochemistry when considering the accumulation or relief of stress through qualities of relationship. There is great value in his approach to wellbeing, and no conflict with the understanding built up in this book; however, his use of the term 'heart' started more in the physical ecology than the way I start to use the term 'inner heart' in this book. We are probably saying much the same things with different emphases.

Part I: The Knowing Brain and the Known World

Chapter 1: Orientated and Aware

1 I am aware that the term 'mindfulness' may be understood in different ways. The exploration of that in New Science terms will be a significant theme throughout this book.

2 Except white, which can be mixed only from a spectrum of *radiant light*. Pigments absorb radiant light, before releasing it again transformed by their own inner molecular ecology. This experiential feature of absorption and inner transformation before releasing into wider life is coherent with the proposed triquetral ecology active at every level of the cosmos.

3 Gerald A. Cory Jr and Russell Gardner Jr (Eds), *The Evolutionary Neuroethology of Paul MacLean: Convergences and Frontiers*. Praeger, Westport, CT & London, 2002. This anthology of papers gives a balanced critique of the strengths and weaknesses of Paul MacLean's original work since the 1970s.

4 I am using the term 'midbrain' in a non-technical sense. Neuroscientists would restrict its reference to a smaller area than I intend to mean by this term. My intention is to give a concept of a highly complex subcortical region of the brain that has far more importance for consciousness than is popularly assumed.

5 Much work has gone into understanding how the thalamo-cortical feedback loops help to bundle together diverse experiences. There are many other arrays of feedback loops within the cerebral hemispheres connecting their deeper structures with other parts of the core brain and limbic systems, adding to the feedback complexity of the inner life.

Chapter 2: An Informational View of Personhood

1 Nouns are *stative,* not *static.* A noun is misconstrued as static if people think there is no change within it. The word 'stative' implies that, despite all appearances of stillness, there is a dynamic process ongoing within. Stative is the 'state of stillness' of a dynamic system, which gives objects their potential to change and respond. Its opposite is 'active'.

2 S. C. Hayes, D. Barnes-Holmes, & B. Roche (Eds), *Relational Frame Theory: A Post-Skinnerian Account of Human Language and Cognition*. Kluwer Academic, Boston & New York, 2002. Steven Hayes is a behavioural psychologist who developed Acceptance and Commitment Therapy as a means to bring covert conditioning into behavioural therapy. The language theory has much wider relevance than its application in any type of therapy.

3 But please do not forget that there are traumatising memories also, which can remain as associations that produce 'transference' of behavioural reactions onto other people with assumptions made inappropriately about

them or their values. Such is the nature of the human inner heart.

4 The endoderm inner lining of the tube becomes the gut and digestive organs with its own internal nervous system (the autonomic nervous system). The mesoderm in the middle becomes bones, connective tissue, muscles, lungs, heart. In this layer, the circulating 'life blood' distributes nutrients and special hormonal communications around the whole body. This is the equivalent communication 'network' that unites the mesoderm in a way that integrates its functioning relationally with the two nervous systems in endoderm and ectoderm. Together they make a whole organism emerging, gathering, and reacting in its environmental ecology.

5 Medically, after a stroke has disabled one side of the body and rendered it insensitive, recovering patients sometimes do not recognise their affected limbs as their own. The arm is just an object that is on the bed beside 'me'.

6 John Zizioulas, *Being as Communion: Studies in Personhood and the Church*. St Vladimir's Seminary Press, Crestwood, NY, 1997.

7 An icon is a visual representation of a hidden reality. A symbol is a visual representation of an idea.

8 A Mexican Wave is an effect moving through a crowded sports stadium that resembles a moving wave produced by successive sections of the crowd in a stadium standing up, raising their arms, lowering them, and sitting down again. It happens because of complex subtle communication patterns among people with similar personal values.

9 A phase space is a 3D physical space that becomes a 4D movement space by adding time. The triquetra is an idealised 4D phase space. It shows iconically in principle how diversity and unpredictability can arise within the integrated movements of its central uniting zone. Variability is created by modulating the relational inputs

at the three corners, resulting in adaptable intra-actions between the three independent elements of integrated movement. These are the necessary and sufficient conditions for indeterminacy to emerge from within adaptive systems.

10 David J. Griffiths, *Introduction to Quantum Mechanics* (2nd Edition). Prentice Hall, 2004, p. 311.

Chapter 3: Becoming Informationally Orientated

1 The drugs used to control attention deficit hyperactivity disorder (ADHD) are paradoxically stimulants, because they stimulate the *inhibitory* synapse-on-synapse filters.

2 Although in Boolean algebra these relational states are confusingly also called operators.

3 Thomas Kuhn, *The Structure of Scientific Revolutions.* University of Chicago Press, 1962.

4 I use the words 'cosmos' and 'cosmology' when I want to emphasise how mindfulness is an integral feature of the material universe. I use the word 'universe' when I want to emphasise the physical ecology out of which life-forms emerge. Both terms, as I use them, imply a view of matter not as dead, but as inherently in informational movement that makes the emergence of life almost inevitable where the physical conditions are diverse and stable enough.

5 https://plato.stanford.edu/entries/process-philosophy/ October 2017, last viewed 24 December 2021.

6 Elizabeth Grosz, 'Bergson, Deleuze and the Becoming of Unbecoming.' *Parallax*, 11:2, 4–13. https://doi.org/10.1080/13534640500058434

7 The dynamic shown in the ancient Celtic icon of the triquetra visually represents equally well the Buddhist concept of interdependent co-origination (Sanskrit, *pratityasamutpada*) and the Christian theological notion of triunity. It has therefore been repeatedly discovered in

different intellectual contexts, the most recent of which may now be quantum cosmology science.

8 Carlo Rovelli, *Reality Is Not What It Seems: The Journey to Quantum Gravity*, 2014. Penguin Random House, 2017. Translated by Simon Carnell and Erica Segre, 2016.

9 J. Gleick, *Chaos: Making a New Science* (20th Anniversary Edition). Penguin, 2008.

10 P. Cilliers, *Complexity and Postmodernism: Understanding Complex Systems*. Routledge, 1998.

Part II: Heart-Level Emotion Is Energy in Motion

Chapter 4: Mind as Consciousness, Thought, Will, and Emotion

1 A. Kring and D. Sloan (Eds), *Emotion Regulation and Psychopathology: A Transdiagnostic Approach to Etiology and Treatment*, Guildford Press, New York, London, 2010.

2 Kevin R. Murphy (Ed.), *A Critique of Emotional Intelligence: What Are the Problems and How Can They Be Fixed?* Psychology Press, 2014.

3 A. Damasio, *Descartes' Error: Emotion, Reason and the Human Brain*. Papermac, 1996.

4 Perhaps the most relevant to the theme of this book is: A. Damasio, *The Feeling of What Happens: Body, Emotion and the Making of Consciousness*. William Heinemann, London, 2000.

5 Distortions of body image are involved in various mental illnesses.

Chapter 5: Emotional Chaos Theory and Emergent Order

1 Core limbic brain activation of physical emotions is the core survival origin of emotional experience as *social physiology*. When people reactivate emotions from recalled

memories, however, the process moves in the opposite direction informationally into the body physiology. In the Emotional Logic system this would be called 'feelings of emotion'. Social physiology *physical emotions* are the real-time evidence of challenged personal values in a social ecology. Feelings of emotion are the evidence of past values, which can trap people in their past and prevent exploratory movement forward in life.

2 https://tools.emotionallogiccentre.org.uk/card-patterns/demo Last accessed 15 May 2022. This is a free 'sandbox' version for exploration only that does not save patterns, but also lacks full instructions. A fuller introductory description can be purchased at low cost from: https://emotionallogicshop.company.site/Online-Introduction-Course-p289896609

3 By convention, capital letters are used for the names of the emotional Stepping Stones because they are proper nouns describing inner physiological states. When referring to fleeting informational feelings, lower-case initials are used.

4 T. Griffiths & M. Langsford, *Emotional Logic: Harnessing Your Emotions into Inner Strength*. Hammersmith Health Books, 2021. This casebook of learning interventions shows how eight common whirlpools of emotion have been identified that are associated with common mental illnesses and socially disruptive behaviour, all of which can improve on learning the Emotional Logic of healthy adaptability.

5 https://emotionallogiccentre.org.uk, or elcentre.org

Chapter 6: Energy and Informational Processes Are 'Informenergy'

1 Mark Carney, *Value(s): Building a Better World for All*. William Collins, London, 2021.

2 J. Ballatt & P. Campling, *Intelligent Kindness: Reforming the Culture of Healthcare*. RCPsych Publications, London, 2011.

3 Some would say there has been a heavy price to pay, in terms of mental or psychosocial health, to achieve material comfort and economic performance.

4 Alfred North Whitehead, *Process and Reality*. The Free Press, New York, 1978.

5 Change is only knowable as the dynamic relatedness of forms; relatedness is only knowable as the mutual change of forms; form is only knowable in the boundaried changes of relatedness.

6 elcentre.org offers the 'front end' practical way forward after recognising that this 'back end' philosophy can liberate life into exploration based on named personal values.

7 David Bohm, *Wholeness and the Implicate Order*. Routledge, 1980.

Chapter 7: A Place for the Human Heart in Physical Science

1 *Psyche*, soul-life, was thought to be emergent as breath, heartbeat, and foetal movements in the womb (quickening). These are cyclical movements emerging within a living creature. *Pneuma* is likened to the wind passing through the ecology. They are importantly different features of life, in that the breathing movements are evidence of life within, while the special feature of the wind is that no-one knows where it comes from or where it goes to, but it is known only by its effects locally on the movement of life.

2 Aristotle described four types of cause: material (properties of substance), formal (design), final (purpose), and efficient (movement, or how). Enlightenment Science method excludes all purpose and design from experiments on the world *other than the experimenter's*.

Part III: Conversations between People Who See Life Differently

Chapter 8: Three Analytical Perspectives Can Disrupt Conversations

1 Martin Buber, *I and Thou*, 1923. Translated and published in English by Charles Scribner's Sons in 1937. Reprinted 2004 by Continuum, ISBN 978-0-8264-7693-7.

2 Applying the previously outlined types of cause to this example: material causes—the qualities of the materials used affect the process; formal causes—the form of a boat in Aristotelian thought is its design category of things that look like boats, for example, not just a plank in the water; efficient causes—the qualities of the tools that will need to be used; final causes—the design purpose: is it fit for use?

Chapter 9: Other Variants of Triquetral Diversity in Unity

1 A true dualism of incompatible essences is found in Persian Zoroastrianism.

2 As will be described in Part IV, the triquetral quantum change could start as an emergent circle from a point, which folds into left or right spins as it develops potential entanglements in an ecology of movement. When synergically entangled, a bidirectional mutuality of informational flow creates existence that can scale fractally through further entanglements to propagate energy flow. Left and right spins innately diversify the movements that emerge within this uniting ecology of life.

3 This confusion of substituting tripartite thinking for triune life relatedness may have had an unfortunate long-term consequence. There is a Venn diagram in the floor of the Hagia Sophia cathedral (now a mosque, having been a

museum) in Istanbul, built in the sixth century CE. This suggests that some Early Church theologians were tripartite thinkers, to which the Prophet Muhammad objected that Christianity was proclaiming three gods. The resulting religion of Islam firmly asserts One God, as does the Jewish faith, which would never have been such a divisive issue had the pure triquetra been used in that centre of Byzantine Christianity to represent One Living God manifest in three utterly relational persons (substates).

Chapter 10: Speaking Life

1 Mahayana Buddhism is therefore more triquetrally balanced than Theravada Buddhism.

Part IV: Belonging Together in a Land of Quantum Gravity

Chapter 11: Movement Shapes a Quantum Gravity Universe

1 The dot points may be reminiscent of a proposed concept once called a *preon*. This Zen paradox adds to preons the process of triquetral change, which is their emerging potential for informational movement.

Chapter 12: Movement Shapes Our Social Ecology

1 https://www.youtube.com/watch?v=X42oZ1O8hAk&t=1s&ab_channel=WWESChannel, https://www.youtube.com/watch?v=FN325bUhCEw&ab_channel=xstreamprayerlive

2 T. Beauchamp & J. Childress, *Principles of Biomedical Ethics* (7th Edition). Oxford University Press, 2013.

Glossary

Adaptive systems. A system is a collection of independent elements in movement, each of which has a rich pattern of bidirectional communications with others creating feedback loops of change that may become stabilised but may also vary with learning. The informational feedback from changes in remote areas of the system makes the pattern of communications within the whole system adaptable when wider environmental conditions change.

Anticipation pattern. Sensory inputs to the central nervous system have a synaptic relay in the core brain between individual neurones that are connected into wider networks (subsystems). Inputs from higher in the central nervous system (CNS) vary the filtering activity of these synapses. In this way the anticipated outcomes of planned actions can prime selection of relevant returning sensory input, contributing to attention focusing and goal-directed activity. The patterning embedded continuously into this filtering is called an anticipation pattern. Habitual preferences in anticipations may be saved in the cerebellum to induce bias into behaviour that may be difficult to relearn.

Atomism. The philosophical belief that reality is constructed from small 'building blocks' called atoms that are self-existent. In more general terms, it is the belief that anything can be analysed into distinct, separable, and independent elementary components.

Attention. The focusing of informational processing on features of life that are personally interesting, while suppressing sensitivity and responsiveness to other dynamic features.

Attractor state. The habitual communication pattern of an adaptive system that has become stabilised and less responsive, or unresponsive, to wider ecological changing conditions. New informenergy inputs to a system in its attractor state become repatterned in that same way. Vibrational informenergy can accumulate within the overall attractor state until it diversifies and becomes unstable, when the whole system will bifurcate into different states.

Awareness. The state of a self-organising living system where sensitivities to change and responsiveness formulate a self-identity that is not orientated in an ecology. This process of self-emergence in a human being occurs below the orientation horizon. It is a prerequisite for comparative consciousness.

Butterfly Effect. A small change in the communication patterning of an adaptive system that has a large effect on the way the whole system processes and impacts surrounding environments. This scaling of impact is called a Butterfly Effect following the original description that a butterfly flapping its wings can influence the course of a tornado.

Cause-effect. Movement in a linear time sequence from past to future, where past conditions solely determine the subsequent repatterning of movement. This type of analytical focus enables experimental interventions and measurable outcomes.

Chaos. The state of a system in which there is no or little stability of communication patterns. The behaviour of such a system is unpredictable and may seem to be formless. In diversified zones of a chaotic system, changes of state can occur that restore ordered patterning of communications and seed a spread of renewed order. A helpful metaphor is to liken chaos to steam that can condense into droplets of water when environmental conditions change. Liquid water is metaphorically a state of complexity, which can further simplify into stabilised repatterning as ice.

Character. Those features of the behaviour and lived experience of a person or system that emerge by consciously comparative choice, having learnt from past experience or future anticipations. Character may be contrasted or compared with personality, which refers to features of behaviour that are inherent to the physical patterning of self-organising behaviour within an ecology. A third feature identifiable in behaviour may be called the spirituality of qualities of relatedness.

Closed system. Any system in which there is no energy or information transfer in or out (no informenergy movement). The concept of entropy and the traditional Laws of Thermodynamics apply only in closed systems.

Complexity. The state of a dynamic system in which patterns of bidirectional communication and responsiveness can partially stabilise in different zones or substates as environmental conditions change. Informational flow thus emerges between diversifying features of the system. The capacity to stabilise forms within timeframes gives sufficient consistency in complexity for growth, adaptability, and the potential to self-replicate. Metaphorically, liquid water is in a state of complexity with different levels of dynamic in its substates as the temperature varies and induces extra movement. The complex state can evaporate into steam as it bifurcates into a chaotic state, or solidify into ice as it bifurcates into a simple patterning state. Liquid water allows life to develop and self-replicate, while steam and ice cannot sustain life.

Consciousness. An experience of awareness in a shared ecology arising from the comparison of that aware self-substate with another non-self, whether that other be a physical ecology or a social ecology. The Latin *con-* (with) and *scire* (to know) suggests responsive movement *with another*. Consciousness introduces the comparative and diversifying capacity to awareness that is needed for orientation in space, time, and substance. With this added quality to awareness emerges also the potential to develop a comparative sense of personal presence and agency in wider environments.

Contextuality. The observation in quantum computing that the behaviour of a 'particle' (for example its spin) can change depending upon its environmental context. Contextuality with non-locality are key features found in quantum science that explore the interface of explicate-order concepts of space, time, and substance with implicate-order synchronicity and non-local entanglement in triquetral holomovement.

Continuous creation. The systemic process whereby forms are continuously emerging as dynamic patterns among changes of relatedness in an open system. These forms may be more or less transitory or stabilised. Patterning of forms moves as informenergy in the implicate order of wholeness below the human orientation horizon. In the explicate order of life, as mentally constructed above the human orientation horizon, these life-enhancing implicate movements may produce effects that seem to be beyond the space-time-substance boundaries of mentally constructed order. In the implicate order, which may be discerned with heart-level awareness, continuously created forms may disperse and be reordered in relation to each other without disrupting the overall informenergy of life-enhancing movement. Informenergy movement may even provide an ecological context where movement is relationally induced in nothing, accounting for an accelerating expanding cosmos by continuous creation potentially everywhere. Likewise, black holes may be a context where organised movement is reduced to nothing.

Conversational orientation. The process of mentally constructing, out of experiential sensory and memory phenomena, a conceptual framework of space, time, and substance in which to communicate personal states and stories with others who are similarly conceptualising their physical and social ecologies. This bidirectional movement of informenergy establishes the personal status of mutual understanding that an individual may discover, as a uniquely synergic substate in a shared wholeness.

Core limbic brain. As described by Paul MacLean, this is the central, highly complex area of the human brain that is usefully categorised together to explain its overall functional connectivity, not its detailed neuroanatomy. The 'higher' level is the cerebral cortex, which adds pattern-recognition, analysis, and memorising capacities. The 'lower' level is the brainstem, which coordinates basic life-supporting connectivity in the body such as matching breathing and blood circulation to activity levels, sometimes unhelpfully known as the 'reptilian brain'. The core limbic brain is a complex network of interacting and balancing sensory and motor functions. Emotive social reactions and intrinsic behavioural dispositions mix here to create a dynamic state knowable as our personal values and identity. The cerebellum may have a significant role in the core limbic brain to stabilise values in balancing behavioural dispositions.

Cortical column. A tube of small neurones perpendicular to the surface of the cerebral cortex where inputs are processed to produce modified outputs within feedback loops. Their function is similar to electronic radio valves or transistors, or solid state microprocessors. Their processing role can be recruited into different higher-order circuits by changing their inputs, a process known as neuroplasticity.

Cosmos. The entire universe considered in the ecological view that life-forms are emergent within its physically structured movements. In triquetral cosmology, cosmos is a non-duality process. This process can, however, be neuropsychologically reduced to a dualistic physical universe.

Deathliness. Dis-integration, when the relatedness element is removed from movement at a relevant system scale. This creates a closed energetic system, where the state tends towards entropic sameness throughout with no informational exchange. The closed system may, however, be set within wider and open systems that relate conversationally with an implicate order that is not framed in space, time, and substance.

Deontological ethics. Moral principles for shared living that are duty-based, by which people recognise their subsidiarity as substates within a higher-order system that can sustain life. Deontological ethics may be compared and contrasted with utilitarian ethics.

Determinism. The doctrine that all events, including human action, are ultimately determined by previously existing causes regarded as external to the will. In a Materialist perspective, linear cause-effect determinism additionally includes belief in the predictability of outcomes when known variables are experimentally controlled in a life-situation, any variation from this being attributed to randomness. (See also **Indeterminacy**).

Ecology. The dynamic relations of organisms to one another in their physical environments.

Emotion. Changing physiological states of whole-body chemistry that are genetically part of ecological survival and thriving responses. These whole-person states include patterning brain activity and social messaging through facial expressions, body posture, tone of voice, and pheromone chemical release. Emotions arise reactively as preparations to respond when survival and personal values are relationally challenged. (See also **Feelings of emotion**.)

Emotional Logic. The partnership of reasoning and emotion to understand how a healthy adjustment process to change can be energised when intentionally and rationally focused on preserving or recovering named personal values. The resulting solution-focused action plans intra-actively shape social systems around explicit personal values.

Emotional Stepping Stones. States of organisation of the inner ecology involving physiology, thought, and social messaging that are robust and meaningful enough to enable decisions to be consciously made on how to progress rationally through an emotive adjustment process.

Energy. In a triquetral cosmology, energy is a fourth- or higher-dimensional wave of movement through a synergic spinfoam of triquetral waves. This synergic spinfoam is the implicate-order holomovement that creates the foundational three spatial dimensions. Informenergy movements add a fourth dimension of time to that space, in which comparison of movements creates a concept of measurable energy moving in an explicate-order framework of space-time-substance.

Enlightenment Science. The experimental method sanctioned by Royal Charter from King Charles II in 1662 when founding the Royal Society of London 'whose studies are to be applied to further promoting by the authority of experiments the sciences of natural things and of useful arts'. For 300 years a materialistic, objective experimental method influenced the development of science globally, but in the 1970s digital and analogue computers became more powerful, allowing the development of chaos theory and complex adaptive systems analysis, known as the New Science. Both approaches to scientific study allow comparison and verification of observations made by different teams, with a truth criterion of replicability. However, the New Science of contextual dynamic systems also allows indeterminism, so that probability becomes part of 'scientific truth'. One historical view is that the Royal Charter was granted to prevent charlatans from misleading the population with false claims of esoteric powers that play on people's superstitions. Historically, however, the materialistic approach adopted also led to deism, eventually displacing all spirituality from a materialistic view of the cosmos. Dualistic body-mind thinking became the dominant paradigm. In its turn, dualism gained its own esoteric power over the population through widespread 'Enlightenment science' education. The New Science systemic paradigm, however, now allows a non-dualistic reinterpretation of the material world that can include a spirituality of relatedness in the probabilistic science of life.

Entanglement. A 'spooky effect' of quantum indeterminacy and contextuality is that measurements of quantum change at two different explicate-order localities may be linked in some way through implicate-order holomovement.

Entropy. The lack of order or predictability within a system. A closed thermodynamic system drifts towards entropy (increasing entropy) as order gradually declines into disorder or randomness of movement of its elements. There are mathematical methods to measure the degree of disorder within a defined system, which represents the unavailability of a system's thermal energy for conversion into mechanical work. In Shannon's Information Theory, entropy has a different but related meaning. It refers to the informational impact (surprise) of a specific message as it arrives within a system. That 'surprise' is the inverse probability of that specific variable event occurring in that system. Complete randomness of the system (high entropy) means any order arriving in the system has high information value, which reduces entropy. The 'Shannon information entropy' applies therefore to open systems. It is a mathematical relationship of change between the transmission-arrival event and its probability. In classical thermodynamics, entropy is defined in terms only of macroscopic measurements. It makes no reference to any probability distribution of an incoming change. By contrast, probability is central to measurements of 'Shannon information entropy' in open systems.

Epigenetics. The study of changes in organisms that are caused by modification of their inbuilt genetic expression. Epigenetics resolves the 'nature or nurture' dilemma when trying to explain observable phenomena of life, because changes of gene expression are mostly caused by environmental influences on the whole organism or its subsystems.

Existence. The dynamic presence that changes the relatedness of other patterned forms, especially when sentient others can experience the change in a shared reality. From within consciously constructed different analytical perspectives on life, the qualities of existence will be described differently. Descriptions will vary depending upon whether the primary analytical focus is on the forms, the changes, or the relatedness qualities of the experienced relational dynamics.

Experience. The phenomena that are produced from systemically going through or engaging in a changing situation. This word can also mean the cumulative effect of memorised experiences on learned behaviour, and on the capacity to make wise choices.

Explicate order. A term coined by David Bohm in the 1980s in *Wholeness and the Implicate Order* that differentiates an order of reality at quantum level called the implicate order from an 'explicate order' of human conscious interpretation. The implicate order functions in wholeness prior to notions of space and time. The explicate order differentiates and separates experiences within that wholeness, but in ways that may not truly represent the implicate order's relational integration as 'holomovement'. Triquetral cosmology calls the interface of implicate and explicate orders the 'orientation horizon'. Triquetral cosmology adds to David Bohm's concepts a neurophysiological explanation for how the explicate order is framed also in an experience of substance, constructed from changing qualities of relatedness in the implicate order.

Feedback. Modification of a process or system by its results or effects. Establishing a feedback cycle relies on sensitivity and responsiveness of the elements of a system in their communications.

Feelings of emotion. Feelings of emotion arise when awareness of embodied changes connects with memories of past situational changes. Awareness thus becomes comparatively conscious and reflective. This comparative and analytic process enables understanding and misunderstanding to further influence innate survival responses.

Fission. Splitting of a heavy atomic nucleus with the release of energy.

Fractal pattern. The pattern of variability of a simple feedback cycle within a dynamic system. A stable system has a range of internal variability that adds unpredictability into its sensitivity and responsiveness when interacting with other dynamics and systems. The influence of these multiple feedback cycle patterns and stabilities can scale up to influence the structural shaping of dynamics at higher levels of systemic order. This scaling up of patterns may introduce a degree of deterministic behaviour and predictability into life.

Fusion. Union of atomic nuclei to produce heavier atoms, with release of energy.

Fuzzy boundaries. The zone where two dynamic systems meet, such that their fractal patterns of change influence and modify each other. This boundary or 'liminal zone' is the physical reality of relatedness in the systemic processes of life.

Gap in time. The experience of letting go of the phenomena that have been framed in analytical perspectives. These three primary perspectives make time seem to be a linear sequence, or a timeframe, or the eternal now. Experiencing a gap in time is a contemplative but mindfully engaged state of living, in which open awareness is allowed while synergically active or stative, choosing not to focus attention on explicate-order energies. It is a mental and spiritual state in which to explore sensitivity to and awareness of wider synergies of heart-level holomovement.

Harmonic. The nodes of a vibration along a tensioned string, multiples of which subdivide its natural frequency to give it a complex tone. By lightly touching a vibrating string at one of its nodal points, a pure or simple tone can be produced that may more clearly resonate with another tensioned string and induce movement there. The blending of complex vibrations, such that features of it produce consonance and tonal resonance, generates new integrative qualities of experience called harmony.

Hidden losses. Values that have been challenged by systemic change enough to induce emotive loss reactions, but which have not been consciously named. In a changing situation, people will therefore lack insight into the causes of their experienced loss emotions. Consciously choosing to name their hidden losses, or worries about loss, in a changing situation is an effective way to recognise the personal values that constitute self-identity, which empowers more constructive action-planning.

Holomovement. David Bohm's term referring to the notion that, at a quantum level, the cosmos has an implicate order that is a united whole in constant process. The human being participates in this process more or less consciously.

Homunculus. A dualistic notion that the human person has a self-existent soul or presence that can interact in an influential way with the physical order of life. In medical or neuroscience terminology, however, the homunculus has a different meaning, as a map of the body distributed spatially across the surface of the parietal lobe sensory cerebral cortex.

Implicate order. See **Explicate order** for an explanation of how these two terms define each other.

Indeterminacy. In philosophy, indeterminacy is the uncertainty that arises when not all causal factors in an existing situation are known. The notion of randomness may then be invoked to explain the reduced predictability of outcomes. In quantum physics, however, indeterminacy more fundamentally means that the state of the situation or system, upon which any contextual factors have an effect, is not yet intrinsically formed. The form of movement is emergent from the entire *contextuality* or ecological setting of changing relatedness. In a quantum-context experiment, the outcomes vary unpredictably as if the form at the centre of attention had already been either a wave or a particle. Indeterminacy of the wave-particle prior state is thus not the same as uncertainty of outcome from a known state.

Informational movement. A triquetral cosmology qualitative concept that differs from the Shannon concept of quantifiable, unidirectional information transfer. It is a systems concept, where substates of the implicate-order wholeness mutually change in their patterned forms of relatedness relative to each other at their fuzzy boundaries. In this way, one substate informs another of its changes 'remotely' in an explicate order as they emerge.

Information Theory. The mathematics of a wide range of information technologies, all of which use the Shannon concept of quantifiable information that transfers unidirectionally from a transmitter into a receiving system.

Informationist. The dualistic view that may emerge when someone habitually narrows their mental focus on life into an Individualist analytical perspective (as a Creator of life). In this view, information gains the quality of a self-existent essence, the ordering of which constructs life. Quantifiable information is seen as objectively existent, perhaps geometrically shaped, and requiring energy to transmit. This contrasts with triquetral cosmology in which information is a verb referring to a mutually synergic process of conversational connection.

Informenergy. A non-dualistic concept in a triquetral cosmology that unites the physics of energy with the mathematics of information technology. It achieves this union by describing implicate-order holomovement as the qualitative substrate through which quantifiable explicate-order energy and information seems to move.

Inner heart. The core of life-enhancing personal values that influences the emotive behaviour of a living creature emerging in its physical and social ecologies. In a triquetral cosmology, personal values have a dynamic informenergy nature that connects with the implicate order. This triquetrally inner organising dynamic runs through the emerging structures of the explicate order of living body-brain-social-messaging whole systems. The potential coherence between an implicate-order informenergy physiology and the explicate-order molecular physiology is the nature of inner heart.

Interdependent co-origination. The concept central to both Buddhism and Christianity that creation or renewal is an emergent process that involves interaction or conversation. The Sanskrit term is *pratityasamutpada*. The Christian theological notion of triunity describes this process as personal substates of one living deity in a conversational unity that sources a diversifying existence. In secular science, a hypothesised quantum implicate order of holomovement is made visible iconically by the Celtic triquetra as an interdependent dynamic of principles of organisation. The explicate order of existence originates in this implicate order of interdependence.

Karma. The notion associated with a reincarnation view of life that the soul carries the consequences of former choices and actions into the next life, affecting destiny.

Kinaesthetic learning. The use of movement and sensations to assist a learning and memorising process. Kinaesthetic learning is very effective for making sense of the useful purposes of emotional states, because feelings of emotion emerge preverbally from a somatosensory background of brain activity.

Life. A continuous state of functional activity that self-organises the physical ecology into recognisable forms within their environments. Dynamic life-forms have the capacity to self-replicate, and to informationally relate to each other for nutrients and other life-enhancing values in cooperative or competitive dynamics. In some life-forms these self-organising and socialising dynamics may be associated with awareness, or further with comparative consciousness, which may generate an inner mental life that further intra-acts with the physical and social ecologies of life.

Liminal zone. Fuzzy boundaries where the states of different systems mingle to generate new movement. This boundary zone of systems is the informational and energetic reality of relatedness.

Logic. The study of correct reasoning to a conclusion or opinion about an outcome of a process, whether material or informational. Logic describes the process that ensures the inferences to reach that conclusion are informed by reasonably justified evidence, opinions, or known facts. Inductive logic starts from specific observations and makes inferences about generalities (the method of Enlightenment Science research). Deductive logic starts from general statements and makes inferences from them about specific situations (the method that generates hypotheses to test, and paradigm shifts in thinking).

Loop quantum gravity. An attempt to develop a quantum theory of gravity based directly on Einstein's geometric formulation rather than treating gravity as a force. The theory postulates that the structure of space and time is composed of finite loops woven into an extremely fine fabric or network, called spin networks or spinfoam.

Loss. The experience of challenge to personal values for survival, thriving, or reproduction following relational change that results in separation from, brokenness of, or misunderstanding about those life-enhancing attractor states.

M-theory. A mathematical formulation that brings together several variants of the string theory of matter and physical forces to construct a view of quantum gravity. The theory requires 11 dimensions to explain the phenomena of life. Integration with triquetral cosmology may be possible if these dimensions are different topological types of vibration.

Materialist. The dualistic view that may emerge when someone habitually narrows their mental focus on life into a Structuralist analytical perspective (as an Observer of life). In this view, energy-matter has the quality of an objectively self-existent essence, the ordering of which constructs life.

Mindfulness. A non-dualist state of mental engagement in the experience of life. Although mental calming may assist engagement in the non-dualist state, mindfulness is commonly mistaught as a set of practices simply to bring about mental calming. The concept of mindfulness originated in the Buddhist notion of inaction to overcome suffering. However, the association of mindfulness with inaction is neither necessary nor sufficient to achieve the non-dualist state of engagement with life. In triquetral cosmology, that state of engagement is associated with exercising the agency of choice amid quantum indeterminacy to shape life's movements and action with others with compassion during times of local context change, or wider ecological change.

Movement. In triquetral cosmology, movement is the changing relatedness of patterned forms from which life may emerge. This systemic concept is the set of principles of organisation that apply at every cosmological level emerging from implicate-order holomovement.

Neocortex. The folded surface of the two cerebral hemispheres a few millimetres in depth in mammals. Several layers of small nerve cells form cortical columns in the neocortex perpendicular to the surface. These cortical columns informationally process the biochemistry and biophysics of neural physiology as components of multiple feedback loops throughout the whole brain.

Neuroplasticity. The neurophysiological concept that cortical columns can be recruited into variable informational purposes as their inputs connect into changing feedback loop patterns of neural activity. Inherent learning through changes of synaptic connections can introduce bias into their connection loops, which can to some extent also be relearnt.

New Science. A dynamic adaptive systems approach to the scientific study of emergent physical, social, and inner ecology phenomena. The method relies on high levels of mathematical computing of simultaneous processes. Algorithms model movement that can be emergent at any systemic level. The method provides probabilistic predictions of behaviour that can be compared also between different research teams for verifiability.

Non-locality. Experimentally inducing a change in one wave-particle can simultaneously alter the measurement of change in another wave-particle measured even miles away. This is one of the 'spooky phenomena' of quantum physics experiments called entanglement, which together illustrate the paradox of explicate-order space, time, and substance in relation to implicate-order dynamics. The nature of this remote entanglement is much in dispute, but it is an undeniable feature of the world we co-inhabit that seems mostly well ordered and predictable.

Object recognition. The neurological process occurring in the temporal lobe sensory association area generating a mental impression that knowable objects have a continuing or unchanging existence. In the triquetral cosmology hypothesis of personal intra-action, the temporal lobe sensory association area would integrate change and relatedness phenomena from sensory inputs and memories to construct object recognition. This synthetic process conceals the relationality of apparently separate objects.

Open system. An adaptive system where the fuzzy boundary liminal zones allow informational energy transfer in and out, as relational intra-action in ecologies.

Orientation horizon. The construction of a stabilising mental framework of space, time, and substance for conversational orientation in ecologies. This mental horizon is an attractor state for informational processing around personal values and agency. Personal awareness of self in a physical and social ecology is formulated prior to, or below, the construction of this mental orientation horizon. Above the orientation horizon in space, time, and substance, the quality of orientating mental framework can be altered by adopting analytical perspectives in this conversational quality of consciousness. The analytical perspectives have a purpose of improving understanding and agency, but if any perspective becomes a habitual view of life there is a potential for pathologies to narrow or disturb human-orientated experience.

Paradigm. A typical example or pattern of something. In relation to an understanding of life, a paradigm is the patterning of a commonly held worldview that underlies the theories and practices of a particular field of study, such as a scientific view of life. A paradigm shift in science is a change of worldview that requires a reinterpretation of all the previously held beliefs that underlie the theories and practices of that science, probably affecting also other related disciplines of thinking and feeling.

Patterned form. The notion that the form of knowable objects or dynamics, which may seem constant or static, is generated from the consistent fractal patterning of informenergy in the dynamics of its implicate level of order.

Person. A dynamic substate of a greater whole having the living capacity to make informed choices of behaviour or communication by sensitivity to changing ecologies. A person is the accumulation of their values, which are dynamic attractor states shaping their lives and responses within changing ecologies.

Personal values. Attractor states of core brain and associated whole-body physiology that are life-enhancing for the person hoping to live in their preferred ecologies. The core brain includes the cerebellum, which may have a role in stabilising the behavioural responses that contribute to values-based selective attention. Personal values are directly related to survival, thriving, and reproduction in the unique individual's life experience. By contrast, aspirational or organisational values are those chosen to which several individuals may give their assent to shape their intra-actions that would sustain consistent ecologies.

Personality. Those innate features of the behaviour and lived experience of a person or a robotic system that are determined by the inherent patterning that physically forms their behaviour. These features may be compared or contrasted with character, which refers to those features of behaviour that emerge by wilful choice or learning from past experience and future anticipations or hopes. A third feature in behaviour may be called spirituality, which can be equated with sensitivity and responsiveness in the wider relational systems in which a person's personality and character are intra-active.

Phase. When the peaks and troughs of two or more vibrations align with each other, they are said to be 'in phase'. When a peak and a trough coincide they are said to be 'out of phase'.

Phase space. A graphical concept that adds time as a fourth dimension to a 3D physical space. A cyclical process may be considered a vibration affecting that whole phase space. Where there are slight variations of a cyclical feedback process, a fractal pattern of the range of variability of behaviour may emerge within that phase space.

Physiology, molecular. The science of life, especially as focused on the biochemistry and biophysics of cells, organs, and whole organisms that are intra-active within physical and social ecologies.

Physiology, triquetral informenergy. The science of life, especially as focused on the quantum gravity level of physics as it emergently moves explicate order into a molecular physiology within its explicate phase spaces.

Positivism. A philosophical system recognising only observable phenomena and unquestionable facts. Positivist philosophies emphasise objectivity, contrasting this with subjective states, pointing out that linguistic problems of interpretation are associated with subjectivity. Positivism tends therefore to be associated with a dualistic or dual-aspect view of mind.

Presence. The self-organising dynamic of a living form that potentially can be sensed and experienced relationally by another. That other would detect a change in their qualities of relatedness to that dynamic, which might draw their attention to the living form, to which living form they might then attribute the quality of 'presence'.

Primary constructs for orientation. The neuropsychology of conscious orientation as an active personal agent frames the phenomena of living in three primary mental constructions conceptually of space, time, and their experiential understanding of substance as the real presence of objects and other people. These conceptualisations are the neurological outputs from three 'sensory association areas' in the cerebral neocortex, which further integrate in the frontal lobe motor-planning areas to construct the conscious orientation needed for personal agency in changing ecologies.

Principles of organisation. In the context of triquetral cosmology, the term 'principles of organisation' refers to the analysis of any movement (or process) into the changing relatedness of patterned forms. How these three features balance to create an adaptive system of movements determines the system's emergent behaviour. The balancing of these principles also determines the extent to which a pattern of behaviour can scale into different system levels, emerging eventually as self-replicating life.

Process philosophy. An approach to philosophy that identifies processes, changes, or shifting relationships as the true elements of the ordinary, everyday real world. Classical ontology since the time of Plato and Aristotle has focused on enduring substances as the underlying world reality, seeing transient processes as ontologically subordinate. Triquetral cosmology adds detail to David Bohm's alternative process concept of reality, as an implicate-order holomovement. Triquetral cosmology describes holomovement as a synergic feedback reality, from the harmonics and resonances within which emerges an explicate order of substances moving in space and time.

Process triunity. A philosophical alternative to monism, dualism, and dual-aspect monism in which the processing of diverse substates within an implicate-order holomovement intrinsically generates more diversity of substates in an overall unity. The principles of organisation that shape the processes of life are triune, because this is both necessary and sufficient to generate unpredictability and critical change of process. 'Process triunity' is therefore a synonym for 'life'.

Psychology. The study of mind, social behaviour, and the inner experience of life. Enlightenment Science has narrowed this study to frame it in a dualistic separation of mental experience from body-brain, trying then to explain how body and mind relate to each other through the brain. An ecological view of the person overcomes this separation, however, with a process triunity view of life that restores the psychology of inner experience to its wider context of social and physical ecologies.

Quantum. The smallest discrete unit of a phenomenon (plural: quanta). In physics this tends to be associated with the measurability of energy, assuming that energy-matter is the self-existent essence of the human experience of life. In a triquetral cosmology of implicate-order holomovement, however, the phenomena of life's movement has a quantum 'change of relational pattern'. Process (informenergy) requires synergic entanglement of quantal changes that influence whichever system level of explicate order is being considered.

Quantum gravity. The attempt in theoretical physics to describe the gravitational field in terms of quantum behaviour. At the time of writing there are two main competing views: M-theory (uniting previous superstring theories) and loop quantum gravity. Neither adequately explain how gravity integrates with physical forces, possibly because 'a quantum of energy' is the core proposition mathematically. Triquetral cosmology provides a relational process hypothesis instead, that gravity is a feature of an implicate-order synergic network of quantum changes (holomovement), through which energy processes are measurable fourth-dimensional movements that can be compared with each other.

Qubit. A quantum computing term referring to a basic unit of quantum information. The computing qubit comprises two classic binary bits that each may be physically realised with a two-state device as 0 or 1, generating a third indeterminate quantum state between them.

Relational frame theory. A neuropsychology term for preconscious mental associations of phenomena, the relational frame that contributes to the unique meaning of a word or experience for an individual. Relational frames are the activation of neural feedback loops associating memorised phenomena with the current focus of mental attention on ecological dynamics.

Resonance. The amplification or reinforcement of a vibration by coherence with another that vibrates with a range that includes its natural frequency or harmonic nodes of that frequency.

Sensory association areas. Three areas of the sensory cerebral cortex where features from different sense modalities are associated with each other and with memories. These associations project forward with feedback loops to the frontal lobe motor-planning areas, where they contribute an orientating sensorium in which the individual prepares their intra-action as a living agent.

Simplicity. The state of a dynamic system in which the patterning of communications between independent elements is restricted in its ability to repattern. This produces a habitual attractor state of movement. The system may appear static or unchanging to an observer from a different or higher systemic level of order. For example, a skater may perceive ice to be static and safely simple to move across, but the state can bifurcate into a liquid water state of complexity when environmental conditions change, for example the weight of the skater through the blade.

Spinfoam. A theoretical model of very small-scale fluctuations of space-time due to quantum changes. This model may describe how the quantum and gravitational scales of cosmology have a common source. At the time of writing, however, there is no widely accepted model of how gravity interacts with the known forces of physics that can be explained by quantum mechanics.

Spirituality. As interpreted in triquetral cosmology, spirituality refers to the relational features of lived experience moving at the different scales of systemic connection that are relevant to individuals and cultures. The term has equal standing alongside 'character' and 'personality' as perspectives on a person's life experience. Spirituality is thus distinguishable from religious belief or practice, such that humanism and atheism may inclusively be considered to have their own relational qualities of spirituality.

Stable domain. A triquetral cosmology concept where alternating l-spin and d-spin triquetral quantum changes entangle in the implicate order prior to the explicate-order imposition of concepts of space, time, and substance. The relational patterns that result stabilise into hexagonal honeycombs, which topologically generate a 3D spatial structure to triquetral holomovement. This implicate order is thus a synergic spatial spinfoam of triquetral waves connecting stable domains, through which the energies of the explicate order can move.

Standard Model of particle physics. The Standard Model is the current best theory scientists have to describe their experimental findings in terms of the most basic building blocks of the universe. It explains how particles called quarks (which make up protons and neutrons) and leptons (which include electrons) could make up all known matter. It is nevertheless incomplete, as it does not explain gravity, dark energy-matter, the accelerating expansion of the universe, or neutrino oscillations. These limitations may be because it is framed in the paradigm of quantum field theory, which itself assumes the fundamental existence of measurable energy quanta, rather than a fundamental bidirectional synergy of holomovement through which energy can unidirectionally move.

State of a system. The relational patterning of mutual changes among the richly communicating elements that constitute a dynamic system. The state has emergent properties of behaviour and of responsiveness to change in relation to other system states.

String theory. A quantum theory of energy-matter that describes single-dimensional strings that vibrate and interact with each other. Several variants of this basic idea have been described that all require hyperdimensional states, beyond the four of space-time in the Standard Model of particle physics, to describe the phenomena of lived experience.

Substance. The etymological source of the word 'substance' is important for understanding its meaning. It comes from the Old Latin and Old French for *under* (sub-) and *stand* (stance). It refers to that which is firm enough to stand upon or to shape life. Its use is commonly restricted to the physical ecology, but in trinitarian theology it may also have a spiritual usage. In the English language, the word 'understand' has the identical conceptual root, but is used exclusively for the mental experience of coherence among ideas that creates a firm enough basis for logical reasoning or making decisions.

Superposition. A controversial early view, as quantum physics was first developing, that particles could be considered to be the cumulative additions of linear wave states at a point, where quantum states could add or subtract from each other. Its weakness is that the mathematics is based on a linear view of wave progressions producing particles. This results in the view that cumulative quantum states could 'dissemble' into different quantum states. The need for a superposition view of particles and waves is removed by adopting a triquetral view of holomovement. In this view, both particles and waves emerge as different patterning of an implicate-order spinfoam.

Synapse. A junction between two nerve cells shaped like a specialised pod where neurotransmitter chemicals are released, allowing unidirectional informenergy transfers within a neural network. A nerve cell may have thousands of synapses on its membrane from other nerve network neurons, and may project to thousands of synapses at its fine dendrite nerve endings. Some synapses activate the cell they connect to, while others have an inhibitory effect.

Synchronicity. A term coined by the analytical psychologist Carl Jung in the 1920s referring to a non-causally linked coincidence of meaningful events, which reinforces the meaningful content of the thoughts or conversational exchange ongoing at the time. The physicist Wolfgang Pauli and Jung together developed the Pauli-Jung conjecture in the 1950s that the physical structure of the universe could allow non-causal connections between events. The wider scientific community considers this possibility to be beyond verifiability by scientific methods. This paradox of connection could be explained by the triquetral synergy interpretation of David Bohm's implicate-order holomovement. Scientifically measurable energy would be a fourth- (and higher-) dimensional movement unidirectionally through it.

Synergy. A vibrational resonance of two-way, responsive communication where any two self-organising dynamic systems are in a mutual exchange that may be described as conversational and mutually supportive. The interaction produces a combined effect greater than the sum of their separate effects.

System, systemic. A system is a flexible set of things, people, cells, molecules, or any other independently movable element that has rich communicating interconnections and feedback. The dynamics of a system produce unique and adaptive emergent patterns of behaviour over time. Systems that are mentally recognised have fuzzy boundaries with other systems, which together become ecologies that exchange informationally and energetically with each other.

Triquetra (plural: triquetrae). A three-cornered (Latin: *tri-*, *quetra*), curved geometric shape that is not a triangle. The triquetra is an early Celtic icon of unending movement behind all that seems constant. A double triquetra is an icon of synergy. The triquetra is mathematically not the same as a trefoil, which is a similarly unending curved coil or knot. However, the trefoil lacks corners where movement turns, or could become a ligand to transmit and receive movement as in a synergic connection with another triquetral movement.

Triquetral wave. An extended synergic 3D string of connected double triquetrae. The reflection of movement synergically extends through the whole length of that string. Triquetral waves have side ligands that can connect with other triquetral waves and with stable domains. This could make a synergic 3D spinfoam as the implicate order of spatially shaped holomovement. Through the synchronicity of this synergic holomovement could emerge the higher-dimensional energies of an explicate order of existence.

TSIPEX model of conscious mind. Triune Systems In Parallel EXchange. Conscious mind, in this model, is the connection of *diversifying capacity* in the cerebral cortex (through pattern recognition and comparison) in an explicate-order to an implicate-order inner heart-level awareness of a *uniting self* embedded in social and physical ecologies. These two levels of implicate/explicate order are each organised using the same triune principles of organisation. Their dynamics are in a parallel exchange (a double triquetra conversation) where they meet at a fuzzy boundary in the core brain and its associated whole-body physiology. The dynamics of this meeting between triune systems may become knowable as personal values and personal identity. However, these dynamics can dissociate, resulting in various pathologies of personal identity. The model relies on differentiating awareness from consciousness, which is the addition of an inner-ecology comparative capacity to awareness.

Utilitarian ethics. Moral principles for shared living that assert the rights of individuals, as subsidiary substates active within a higher-order system, to shape life according to their local and practical needs.

Values. Dynamic attractor states of living systems that promote survival, thriving, or reproduction. Challenge to values induces a systemic response known as a loss reaction that induces a unique personal fractal range of emotive behaviours, feelings, and thoughts. In this dynamic systems view of the person, values constitute the personal identity. In the spontaneity of active personal life, personal values therefore commonly remain unnamed and unrecognised until challenged, the loss emotions then being the evidence of their existence.

Vitalist. The dualistic view that may emerge when someone habitually narrows their mental focus on life into a Collectivist analytical perspective (as a Participator in life). In this view, an esoteric life-force behind matter has the quality of an objectively self-existent essence or energy, the ordering of which constructs life.

Whirlpool of loss emotions. The maladaptive consequence in a person's feelings and internally driven behaviour when they narrow their emotional range of response to accumulating losses, and to potential loss, from the potential healthy seven states to just two states. The body chemistry and associated social messages and thoughts turn from adaptive rational adjustments into inner drives that seem out of character, which may be diagnosed as common mental illnesses or socially disruptive behaviour.

Yes-and-no-no analysis. The capacity that a person gains when they understand and can recognise the three primary analytical perspectives on life (Structuralist/Materialist; Individualist/Informationist; Collectivist/Vitalist), which enables an analysis of their own or someone else's expressed worldview in terms of the proportional contribution from each of these perspectives. Yes-and-no-no analysis can then inform creative conversations, in which people explore how to develop the flexibility to see life differently from these perspectives, and to integrate these differences, respectfully agreeing on how they may enrich each other's lives.

From the Author

Thank you for buying *Clearing a Way: Unveiling the Mental Tricks That Hide Reality*. You may, as other readers have found, need time to assimilate its paradigm shift of vision into a renewed enjoyment of re-emergent life and its potential fulness. Human beings are not merely bouncing randomly around in a material universe. We belong in a cosmos that makes sense, in which mutually supportive groups make sense.

I am available for discussion and questioning. I hope you will explore this book more deeply in reading groups, where you could clarify how to recognise and move flexibly between the different analytical perspectives on life. My hope is to inspire people to greater inner strength and adaptability. With teleconferencing it is easy for me to meet with groups in webinars, or to meet in person via conference talk discussions.

To contact me, please go through my blog and website www.relatedness.net/, and sign up to follow new Insights and publications. If you have a few moments, please feel free to add your review of the book to your favourite online site for feedback.

The Emotional Logic Centre runs completely independently of me. My calling is to open people's eyes to their potential to adapt together with hope, action, and dignity when environments crumble. I am pleased to be able to confidently point readers to elcentre.org to explore solutions more deeply if facing challenging situations, or to develop personal strengths in advance to face change constructively. You can also learn more about Emotional Logic by reading the casebook, *Emotional Logic: Harnessing Your Emotions into Inner Strength*, or for parents of young children the *Shelly and Friends* series of illustrated children's story books. People have found book reading groups helpful to apply new ways to make sense of emotions in home and work life.

In sincere hope for your future.

Trevor Griffiths

O-BOOKS

SPIRITUALITY

O is a symbol of the world, of oneness and unity; this eye represents knowledge and insight. We publish titles on general spirituality and living a spiritual life. We aim to inform and help you on your own journey in this life.
If you have enjoyed this book, why not tell other readers by posting a review on your preferred book site?

Recent bestsellers from O-Books are:

Heart of Tantric Sex
Diana Richardson
Revealing Eastern secrets of deep love and intimacy to Western couples.
Paperback: 978-1-90381-637-0 ebook: 978-1-84694-637-0

Crystal Prescriptions
The A-Z guide to over 1,200 symptoms and their healing crystals
Judy Hall
The first in the popular series of eight books, this handy little guide is packed as tight as a pill bottle with crystal remedies for ailments.
Paperback: 978-1-90504-740-6 ebook: 978-1-84694-629-5

Shine On

David Ditchfield and J S Jones

What if the aftereffects of a near-death experience were undeniable? What if a person could suddenly produce high-quality paintings of the afterlife, or if they acquired the ability to compose classical symphonies? Meet: David Ditchfield.

Paperback: 978-1-78904-365-5 ebook: 978-1-78904-366-2

The Way of Reiki

The Inner Teachings of Mikao Usui

Frans Stiene

The roadmap for deepening your understanding of the system of Reiki and rediscovering your True Self.

Paperback: 978-1-78535-665-0 ebook: 978-1-78535-744-2

You Are Not Your Thoughts.

Frances Trussell

The journey to a mindful way of being, for those who want to truly know the power of mindfulness.

Paperback: 978-1-78535-816-6 ebook: 978-1-78535-817-3

The Mysteries of the Twelfth Astrological House

Fallen Angels

Carmen Turner-Schott, MSW, LISW

Everyone wants to know more about the most misunderstood house in astrology — the twelfth astrological house.

Paperback: 978-1-78099-343-0 ebook: 978-1-78099-344-7

WhatsApps from Heaven

Louise Hamlin

An account of a bereavement and the extraordinary signs — including WhatsApps — that a retired law lecturer received from her deceased husband.

Paperback: 978-1-78904-947-3 ebook: 978-1-78904-948-0

The Holistic Guide to Your Health & Wellbeing Today

Oliver Rolfe

A holistic guide to improving your complete health, both inside and out.

Paperback: 978-1-78535-392-5 ebook: 978-1-78535-393-2

Cool Sex

Diana Richardson and Wendy Doeleman

For deeply satisfying sex, the real secret is to reduce the heat, to cool down. Discover the empowerment and fulfilment of sex with loving mindfulness.

Paperback: 978-1-78904-351-8 ebook: 978-1-78904-352-5

Creating Real Happiness A to Z

Stephani Grace

Creating Real Happiness A to Z will help you understand the truth that you are not your ego (conditioned self).

Paperback: 978-1-78904-951-0 ebook: 978-1-78904-952-7

A Colourful Dose of Optimism
Jules Standish
It's time for us to look on the bright side, by boosting our mood and lifting our spirit, both in our interiors, as well as in our closet.
Paperback: 978-1-78904-927-5 ebook: 978-1-78904-928-2

Readers of ebooks can buy or view any of these bestsellers by clicking on the live link in the title. Most titles are published in paperback and as an ebook. Paperbacks are available in traditional bookshops. Both print and ebook formats are available online.

Find more titles and sign up to our readers' newsletter at
www.o-books.com

Follow O books on Facebook at **O-books**

For video content, author interviews and more, please subscribe to our YouTube channel:

O-BOOKS Presents

Follow us on social media for book news, promotions and more:

Facebook: O-Books

Instagram: @o_books_mbs

Twitter: @obooks

Tik Tok: @ObooksMBS

www.o-books.com